ROCKING ALASKA

STORIES FROM A FIELD GEOLOGIST

STEVEN A. FECHNER

Rocking Alaska:
Stories From a Field Geolgist

Copyright ©2019 Steven A. Fechner

ISBN: 978-0998547770

FLEETING EDGE PRESS, 2019

Dedication:

To Mary and all U.S. Bureau of Mines employees

This is a map of Alaska showing various geographic features and mining districts.

Bodies of water and regions:
CHUKCHI SEA
BEAUFORT SEA
ARCTIC OCEAN
ARCTIC CIRCLE

Canadian regions:
CANADA
YUKON TERRITORY
NORTHWEST TERRITORIES
BRITISH COLUMBIA

Alaska label:
ALASKA

Mountain ranges:
BROOKS RANGE
ALASKA RANGE

Mining districts and project areas:
Colville Mining District
WGM Project Area
Yentna Mining District
Valdez Creek Mining District
RARE II Study Area

Cities and towns:
Wainwright
Barrow
Prudhoe Bay
Gambell
Hooper Bay
Nome
Kotzebue
Selawik
Bettles
Coldfoot
Unalakleet
Goodnews Bay
Bethel
Galena
Holy Cross
Dillingham
King Salmon
Nenana
Kantishna
Anchorage
Fairbanks
Tok
Circle
Palmer
Seward
Homer
Kenai
Valdez
Glennallen
Haines
Skagway
Juneau
Wrangell

Islands and peninsulas:
SEWARD PENINSULA
Matthew Is
Nunivak Is

Mountains:
Mt McKinley 6,194 m
Mt St Elias 5,489 m
Mt Fairweather 4,671 m

Other features:
Yukon (river)
White Mountains
Fortymile River

Contents

Acknowledgments

I would like to thank all of the people who made this book possible. My wife, Mary, helped to edit the rough drafts. I received valuable comments from Russ Fields, Bill Pennell and Opal Adams. Mike Balen, Jeff Foley, Denise Herzog, Robert Hoekzema, Joseph Kurtak, Martin Maricle, Mark Meyer, Nathan Rathbun and Gary Sherman gave me permission to include them in the book. I would also like to acknowledge everyone who was only mentioned either by their first names or their jobs. I especially would like to acknowledge the contribution of Gary Sherman, who without him, this book would have never been published. Finally, I would like to acknowledge all of the people I worked with in Alaska. It was because of them that I was able to successfully work in Alaska.

Preface

This book is about geological fieldwork in Alaska. It could portray any field geologist's work because anyone who works or has worked in the field in Alaska has had similar experiences. I lived in Alaska and was involved with geological fieldwork in the 1970s, 80s and 90s. The book doesn't describe everything I did while working in Alaska but only tries to depict the life of a ground pounder who was involved in living in the Alaskan bush and evaluating mineral properties. This book, however, would be rather dull if it just talked about fieldwork, so I've tried to interject my personal experiences and stories. These stories are how I remember them. It is important to note that anyone else who was associated with these stories might and probably does remember them differently.

A friend asked me if this book was about Alaskan stories or was it a how-to book. I told him yes. The books I find most interesting are non-fiction books that not only tell what happened to people but also what people did when confronted with certain circumstances. So, if you, the reader, are a geologist you might find this book informative on how fieldwork is accomplished. If you are not a geologist, you might just like the stories.

Persons mentioned by their full name in the book have either given their permission to be in the book or are dead. I apologize to anyone I didn't mention by name, who might have given me permission to use their names if I could have tracked them down.

"A mine is a hole in the ground owned by a liar"
Mark Twain

Columbia Glacier

Chapter 1

The Adventure Begins

My eyes popped open. For a moment I was disoriented. I looked around the room. I was crammed into a lower bunk, my duffel bag with all of my clothes and field gear was next to the head of the bunk and sunlight streamed into the room. I grabbed my analog watch and the hands pointed at 7 and 12. Yipes, 7 o'clock. It looked like I was going to be late for my first day working for Watts, Griffis and McOuat, Inc. (WGM) in Anchorage, Alaska. I grabbed my Dopp kit, towel, and clothes and padded to the bathroom. I couldn't believe the other 7 people who were living in the apartment weren't stirring. I guessed they had probably partied until sunrise the night before. But, that was OK with me. I could take a shower, get dressed, and get something to eat in relative peace and quiet. Luckily I didn't have to spend time shaving because I wouldn't be doing that for the next 4 months. I'd then be ready to face my second day in Alaska before the other seasonal geologists started scrambling to use the facilities. The water felt good poring over my 6-foot 6-inch frame. The hot water helped ease the kinks of sleeping in a 6-foot long bunk bed. The water also helped clear my head of the cobwebs created by traveling from Ft. Collins, Colorado to Alaska, then staying up late swapping "war" stories with the other seasonal geologists who were living in the apartment complex owned by WGM. After drying off and getting dressed, I padded back to my bunk, still wondering why no one else was stirring. After sitting on my bunk, getting ready to put on my shoes, I grabbed my watch again and saw that one hand was pointed at the 12 and the other at the 1. My scientifically-trained mind decided 12:05 didn't make sense. I then put on my glasses and found out I had been reading the watch upside down and it was only 6:30. Ah, the mystery of the sleepy-heads solved. I had been fooled by my internal clock, which was still on Colorado time, the sun position high in the sky and my poor eyesight. What do I do for over an hour now that I am totally awake? I decided to think about how I got to Alaska and what is Alaska. Little did I know at the time this would be the start of 18 years of fieldwork in this wonderful State.

I am a geologist, but of course, I wasn't always a geologist. I was born and raised in the suburbs of Sacramento, California. My boyhood home was in a housing tract built in the early 1950s to accommodate the upcoming boomer generation. It was near enough to the American River so I could spend time in largely undeveloped surroundings away from the housing developments. I had the typical post-World War II childhood. I was the middle child in a middle-class family. There were mostly six of us, which included my parents, older brother, younger sister, myself and a number of dogs. My dad went to work each day at McClellan Air Force Base and my mom stayed at home with us kids until we were old enough to look after ourselves. Then she got a part-time job working for the State of California. I walked by myself to elementary and middle school and took the bus to high school. My parents would have just shaken their heads at the present-day "helicopter" parents because they weren't the hovering types. They let us be pretty independent knowing that their spies, aka neighbors, would rat us out if we caused any kind of trouble.

I grew up loving the outdoors. I guess I can blame my parents who constantly told me to go outside and play; therefore, I spent a lot of time exploring drainage ditches, making forts and playing in the local parks with all of the other neighborhood kids. My love for the natural world was reinforced by my parents who would load us up in the station wagon and head off to the mountains whenever they had time off. Most summers and winters they would rent a cabin either at Donner or Echo Lake, near Lake Tahoe where we'd spend weeks, swimming, fishing, hiking, horseback riding, skiing and sledding. Other times they would pack us all into our trusty station wagon and tour the back roads, national parks and forests of Oregon, Washington, Idaho, Nevada and Montana. One of my first brushes with the geology profession was the time my dad took our family to an area outside of Reno to look at an oil well. He had invested some of his hard-earned money on what turned out to be a duster. I learned from this visit that a person should understand the geology of a mineral property before putting any money into that kind of venture. I also learned Mark Twain was probably speaking from experience when he described a mine as "a hole in the ground owned by a liar." It seems that sentiment even applies to oil wells. Besides spending time in the mountains with my parents, a 4-year stint with the Boy Scouts honed my outdoor skills and reinforced being at ease in nature.

Although my elementary, middle school and high school

educations were markedly unremarkable, one thing I realized during those formative years was I wanted to be some kind of scientist who wasn't going to be stuck in a laboratory or riding a desk. However, it wasn't until I went to American River Community College (AR) that I zeroed in on geology as a career after taking some basic geology courses. I completed all of the basic geology and science courses at AR in 2 years, then transferred to California State University, Chico. It was there I received a Bachelor's degree in geology. I then went to Colorado State University (CSU) in Ft. Collins, where I received a Master's degree.

Although geology is one of the oldest professions, there are very few geologists in the world. I contend that most people in the world have never met a geologist. So, what is geology and what does a geologist do? Gē in Greek means earth or any terrestrial body. Logia in Greek means the study of or discourse. Therefore, geology is the study of the earth or any planetary body. Geology is a broad field that encompasses a lot of disciplines and interests. It is similar to other scientific fields where a person is educated in the basics and can specialize depending on what floats their boat, to coin a phrase. If a person is interested in chemistry, there is geochemistry; in physics, there is geophysics; in biology, there is paleontology; economics, there is mineral economics; law, there is mining law; water, there is geohydrology; and in treasure hunting or mapping things, there is field geology. There is almost a geological discipline for any interest a person might want to pursue, whether it be in the laboratory, office or in the out-of-doors.

Although I was exposed to all of the geologic disciplines I mentioned, I gravitated towards field geology because like I said, I didn't want to be stuck working inside.

I was also fortunate the colleges I attended had field-oriented professors who not only provided a text-book education but also an education based on getting out into the field, hiking, mapping and banging on rocks. My formal education was augmented while in school with a field Master's thesis at CSU, summer jobs mapping for the U.S. Geological Survey (USGS) in Wyoming and a seasonal position evaluating a uranium property for a private company in Colorado.

A turning point in my life came while working towards my Master's Degree at CSU. It was there I met an outdoor loving, fun, beautiful coed, Mary O'Connor. She was studying environmental interpretation (park ranger stuff) and loved working in nature. We fell in love and because we had so much in common we decided to tie the knot in December 1976. Prior to marriage, I was content with getting by with summer jobs. Being married motivated me to find a good, hopefully, full-time geology job. A professor once told me early in my career the geology profession was like riding a wave. When the wave was going up, everyone had jobs, but when the wave crested, there were layoffs. This professor would know about this phenomenon because he had been laid off during one of the down times and took a job teaching at a community college. However, a full-time job was hard to find in 1977 because the profession was still in one of those wave troughs and the only field jobs available were summer jobs. Fortunately, Mary and I came of age when it was OK to be poor. We knew if a full-time job didn't come along we could get by with money made from summer jobs. Seeing as I wasn't having any luck finding full-time employment, I interviewed with various companies for part-time work. Fortunately, a Colorado State alumni, working for WGM in Alaska came to the university looking for field people. I interviewed and was offered a summer job starting in May. It was nice WGM paid more than a lot of full-time jobs and would pay my travel expenses. The pay at $1200 per month was more than double what I received working for the USGS.

Working summers apart might seem like a marriage killer to most people. But, Mary and I were used to working summer jobs apart from each other. This would be a pattern that was repeated for many years. Mary was a big help in preparing for working in Alaska because she had hitchhiked to Alaska when she was in college and gave me good advice prior to starting my new adventure. Therefore, in the middle of May 1977, we kissed goodbye before I flew off to Alaska and Mary drove to Flaming Gorge National Recreation Area in Utah to work for the Forest Service.

Kantishna, Denali National Park

Chapter 2

Alaska

Looking at my watch again, it was only 6:45 a.m. One hour and 15 minutes to go before work started. I got up and looked out the window. Boy, it was really light out. The window looked out onto a parking lot that separated the main WGM office from the WGM-owned apartment building. WGM had purchased the building to house field crews and overflow staff. I could see downtown Anchorage and its two tallest buildings, the Captain Cook Hotel and the Hilton. It looked like it was a vibrant city which had sprung back from the devastating 1964 earthquake. I eventually went out to Earthquake Park near the airport, where the power of the earthquake was still evident. I walked around gaping at the water pipes sticking out of the ground and the remnants of asphalt streets tilted at 45-degree angles. During the earthquake, the once level ground with 75 homes slid 2000 feet into the ocean making the land look like a moonscape. Of note, today, people have built houses in the Earthquake Park area and it is hard to find remnants of any quake damage.

When I arrived in Alaska, people joked the best thing about Anchorage was it is a half hour from Alaska. However, Anchorage seemed like Alaska to me. It is a city bounded by snow-capped mountains on the east and the ocean waters of Cook Inlet on the west, Turnagain Arm on the south and Knik Arm on the north. Also to the north on a clear day, Denali and Mt. Foraker are visible in all of their glory rising from the tree covered plain. Anchorage has a lot of trees. Moose can be found roaming the neighborhoods. A little distance from the WGM office a person had a pet reindeer in his front yard. The city teems with Native or aboriginal peoples and others who look more comfortable in flannel shirts and boots than in three-piece suits and ties. For a person used to cities with elaborate freeway systems, a city with only one main highway didn't seem to disqualify Anchorage from the ranks of a frontier community.

Lying back down on my bunk, I reached over to my duffel bag, picked up some of the literature I had gotten at the airport and at the Anchorage Visitor's Center and started reading. "Alaska, the Last Frontier" and the "Land of the Midnight Sun." I wondered how factual these marketing slogans were, so I continued reading. Alaska is a vast largely unpopulated area. Exact statistics differ, but they all point to the fact Alaska has about 1 person per square mile. In comparison, the average person per square mile in the continental United States (U.S.) is approximately 87. If one takes into account the majority of Alaska's population is concentrated in a few cities and villages, like Anchorage, Juneau, Fairbanks, and Bethel, then the rest of Alaska is largely a vast unpopulated frontier. Now, I don't know if Alaska is the last of all of the frontiers on earth, but it's a catchy phrase nonetheless.

Since I had only been in Alaska one day and it was only May and not the summer solstice, I wondered about the other slogan, the "Land of the Midnight Sun". I found it is not true for the cities of Juneau and Anchorage. They have between 18 and 19 hours of daylight followed by dusk at the summer solstice in June. One can't see the sun at midnight in these cities. However, I found out after living a couple of days in Anchorage if I stayed up late enough I could watch the sunrise at about 2 a.m., which caused my brain and body to say, "Hey, let's start the day." I had to travel north of Fairbanks to experience a midnight sun, which didn't set from June 1st until the second week of July, essentially 6 weeks. In Utqiaġvik, formerly known as Barrow, on the Beaufort Sea, the sun never sets from early May to the end of July.

My perception after further reading of the literature indicated Alaska's seasons can really be divided into two parts that don't necessarily coincide with what is on the calendar. The seasons in Alaska should be divided by the summer and winter solstices. The summer solstice or the longest day of the year occurs around June 20th. The solstice marks the beginning of summer on the calendar but I contend that in Alaska it marks the end of summer. People might scoff at my perception, but when I worked in the field, I could see the nights slowly getting darker and colder after the solstice. By the end of the field season, I could watch the Aurora Borealis dancing across the night sky.

The first day of winter on the calendar, winter solstice or the shortest day of the year occurs around December 20th. My perception again is this day is not the beginning of winter, but the beginning of summer. The days following the solstice, even though they are cold contain more sunlight, which lifts a person's spirit. Anchorage has 5.5 hours of daylight on the shortest day of the year in December. Fairbanks has 4 hours of daylight and in

Utqiaġvik, the sun never comes up from mid-November until mid-January. Who wouldn't feel better knowing there is going to be more sunlight for the next six months after the winter solstice?

A marketing slogan I didn't see in the literature was "Alaska is Really Really Big". The Alaska Visitors Bureau can use this slogan for free if it wants. When Alaska is superimposed over the continental U.S., it stretches from Florida to California and Minnesota to Texas. Alaskans like to point out if Alaska was split in two, Texas would be the third largest state. Alaska is so big and is so far removed from the rest of the U.S. Alaskans refer to the continental U.S. as the "lower 48", the "lower four dozen" or simply as the "outside". When I got out my map of the U.S. I found the map makers have a tendency to shrink the size of Alaska and place it off the southwest coast of California, so their maps make a nice rectangle. This must be confusing to geographically-challenged individuals when they actually see a map of the world or they find out in order to catch a cruise to Alaska, they have to travel to Seattle or Vancouver, B.C. rather than Los Angeles.

Further reading of the literature made me wonder where I might be working in Alaska. Alaska is geographically diverse. It has four major regions or systems; the Arctic coastal plain, the Rocky Mountain system, the central uplands and lowlands and the Pacific mountain system. Little did I know that by the time I left Alaska, I had worked in each of these systems. The Arctic coastal plain is located in the north and is bordered by the Rocky Mountain system or more commonly known as the Brooks Range to the south and the Beaufort Sea on the north. It is a flat-lying plain, therefore, the name. The plain has lakes and rivers. Vegetation is mostly low-lying bushes and grasses. Underneath the surface of the plain is permafrost, which is perennially frozen ground. Few people live in this area. They are concentrated mostly along the coast, with the largest city being Utqiaġvik, which has a population of over 4,200 people. The large Prudhoe Bay oil field is located on the coastal plain.

The Rocky Mountain system is composed of the Brooks Range and its foothills. The system stretches from the Rocky Mountains in Colorado, through Canada and ends at the Chukchi Sea on Alaska's northwest coast. It has mountains as high as 9,000 feet. Vegetation consists of low shrubs and grasses. This is probably the least inhabited system in Alaska. A large village within the system is Anaktuvuk Pass, which has a population of around 300 people.

The largest system by landmass is the central uplands and lowlands. It is an area between the Rocky Mountain system and the Pacific mountain system. The area contains large rivers, the Yukon, Koyukuk, Tanana and Kuskokwim to name a few, flat plains and some mountainous areas like the Yukon-Tanana uplands and Kuskokwim Mountains. The area consists of spindly trees and shrubs below 3,000 feet and swampy areas in the deltas of the major rivers. There are large cities and villages in this system. Fairbanks is the biggest city with a population of over 30,000; however, the outlying area has 60,000 more people. Bethel is the largest city in the Yukon-Kuskokwim Delta region, with a population of over 6,000 people.

The area with the mildest and wildest climate is the Pacific mountain system. In general, the system extends from the Aleutians on the west, through south-central Alaska, then curves down the Pacific coast to southern California. It contains volcanoes in southwest Alaska, glaciers in the mountains and large rivers like the Copper and Susitna. There are many mountain ranges within this system, including Saint Elias Range, Wrangell Mountains, Kenai Mountains, Chugach Mountains, Talkeetna Mountains, and of course, the Alaska Range. The highest mountain in North America: Denali, formerly known as Mt. McKinley is in the Alaska Range. I said it has the mildest and wildest climate because the mild climate around Anchorage and southeast Alaska, specifically Juneau, is where most of the population of Alaska lives. There are close to 500,000 people who live in this system. This is out of a total population of approximately 740,000. The climate is also the wildest, with the weather in the Aleutians noteworthy for its high winds, rain and snow. Temperatures in the Alaska Range can be 80 degrees Fahrenheit in the summer and minus 80 degrees in the winter. Vegetation consists of large trees in lower elevations. At higher elevations above 3,000 feet, there is brush, bare rock and glaciers.

Accessing Alaska is challenging, to say the least. It is costly because it is so remote from the rest of the U.S. It also lacks a lot of infrastructure. Because Alaska has more coastline than all of the coastline in the continental U.S., boat or ship access to many parts of Alaska has always been important. Its road system is minimal. No roads existed in Alaska until the early 1900s. The main highway (the Alaska Highway) linking Alaska with the continental U.S. wasn't completed until World War II. There are less than 3000 miles of roads in Alaska, not counting local roads. 99% of the roads are in the southeastern part of the State. Access to much of Alaska via motor vehicle is therefore limited to driving local roads around communities, like Juneau, Haines, Fairbanks, Anchorage and Nome or between communities like from Anchorage to Homer and Seward on the Kenai Peninsula or Anchorage and Fairbanks or Valdez and Deadhorse at Prudhoe Bay. Alaska does have a railroad. It runs from Seward on the Kenai Peninsula to Fairbanks. Consequently, because of the lack of roads, the most viable access to most of Alaska is via aircraft.

When the airplane was first introduced to Alaska in the early 1900s, after World War I, it revolutionized travel. People didn't have to wait until winter when the ground was frozen to hitch up their dog teams to get to remote areas of Alaska. Also, people didn't have to wait until breakup (when the ice melts off the rivers) in order to use their boats. They could just hop on a plane and fly someplace. Landing strips were primitive, but the early pilots were good at making use of lakes, gravel bars and beaches until infrastructure could catch up with them. Therefore, Alaskans took to this new form of transportation as a duck takes to water or more appropriately an eagle takes to the air, with the first commercial flight occurring in 1927. One out of 58 people in Alaska are pilots and one out of 59 own their own plane, which makes Alaska the most popular state for flying.

Even though Alaska is huge, its residents think of it as a large village because they are isolated from the rest of the U.S. and it has so few people. When I first got to Alaska I was warned it was likely if I told someone something, everyone in Alaska could know about it in a short amount of time. This was due to math. If one person knows something and tells 10 people and each of those people tells 10 people, then after less than 5 times of this happening, everyone in Alaska knows the information. People warned me about this phenomenon because the last person might hear a distorted version of the original information. It is kind of like the telephone game we played as kids where we would sit in a circle and one person would tell the person next to them something and that person would tell the next person, and so on until it got back to the original teller. The information usually got back in a distorted form. Just imagine this happening with information passed between thousands of people.

Wow, I sure did have a lot of information to absorb. It sounded like Alaska was going to be a great experience, but I had to get moving because my apartment mates were stirring, there was breakfast to be rustled up and work to be done.

East side of the Dalton Highway, Brooks Range

Chapter 3

My Career

Cheechako. That's what my boss, Chris, called me when I reported for work in 1977. A cheechako is an Alaskan or Canadian term for someone who knows nothing about the north countries. I guess Chris got that impression when I told him about waking up at 6:05. However, I couldn't pass myself off as a sourdough after one day in the far north. A sourdough is someone who has embraced the Alaskan lifestyle and is comfortable with living in the "Land of the Midnight Sun" or "The Last Frontier". 1977, would mark the first year of my 18-year transformation from a cheechako to someone who was comfortable living and working in the far north.

After Chris aptly defined me as a cheechako, I asked him to tell me what I was going to be doing for the next few months. He told me I was the project geologist on his field crew, which meant he would tell me what had to be done and I would coordinate the work of all of the other geologists. This was a step up from my previous geology jobs, where I was just a field assistant. Chris was a full-time WGM employee in charge of a five-person geology crew, plus a cook and helicopter pilot. Chris informed me we only had a week to gather the other crew members and get all of the field gear together, so it could be shipped to our first field camp. There was no time to waste. Luckily, one of the crew members had been on a WGM crew the summer before and could show me the ropes about outfitting a field crew. We were going to work west of Fairbanks. WGM had a contract with the Doyon Native Corporation to evaluate the mineral potential of their land in the uplands and lowlands system.

In 1971, Congress passed the Alaska Native Claims Settlement Act (ANCSA). This law was passed to settle native claims to land in Alaska. There have been many discussions about this act and I will not try to discuss them because each side might have valid points. But, what is widely believed by most Alaskans is the Act was passed in order to exploit the huge quantity of oil found at Prudhoe Bay in 1968. Native claims to land in Alaska needed to be settled before the pipeline could be built or litigation might have gone on for years. So, the brightest people in the state were assembled to work out a solution in order to promote economic development and self-determination in the Native community. One aspect of the settlement, which was universally accepted was few people in Alaska wanted a reservation system similar to the one established in the continental U.S. Native groups had the option to choose a reservation but only one, Annette Island Reserve, opted for it. Therefore, instead of reservations, the Act established 12 regional corporations based in Alaska. A 13th corporation was set-up to take into account those natives who had moved out of state. Two hundred village corporations were also established. Each regional corporation could select so much land within their region (Doyon received 12.5 million acres). Nearly $1 billion was transferred to these corporations. A billion dollars was real big money in 1971. Once the corporations were set up, each started prioritizing land for selection. Because the Regional Corporations received the mineral rights to all native lands (regional and village lands), they decided the mineral potential of lands within their region would be one criterion for selection. WGM received a contract from Doyon to evaluate their region over a 4 year period.

During my first week in Alaska, work during the day included gathering all of the available field supplies and purchasing any necessary items. Nights were spent with other field geologists who were arriving and leaving shortly for other summer field projects. All of us lived in a couple of WGM apartments. There were about 20 or 30 of us. Every night after dinner, we would rendezvous on the floor of the living room in one of the apartments and people would share Alaska experiences. One night someone started talking about bears and guns. With the mention of guns, people started unlimbering all of their artillery. Everyone had their favorite weapon, which was some kind of a pistol. People had .357s, .40s and .44s. Some swore by nickel-plated revolvers because they wouldn't rust. Me, a boy from California, naive to the dangers of Alaska, was about the only person who didn't have a gun. I guess I really was a cheechako.

One of the geologists I met sitting around at night was Opal Adams. I reconnected with her when I moved to Nevada 30 years later. She eventually hired me for a temporary job after I retired from the Forest Service. Throughout the years, I have run into a number of the old WGM hands who survived the ups and downs of the mineral industry. Small world.

I didn't spend much time at the WGM office in May because not many businesses want to pay people for sitting around not doing fieldwork. So, after sampling what Anchorage had to offer and packing up all of the field gear, it was off to the field until September. Our WGM crew consisted of Chris, the boss, me, his loyal assistant, 3 other seasonal geologists and a cook. A helicopter and its pilot joined us at our first field camp. Everyone, but Chris, one seasonal, and the cook were new to Alaska, even the helicopter pilot.

After September, I returned to Anchorage. WGM kept me on in order to organize all of the field data collected during the summer. My wife, Mary, joined me. WGM let us stay in one of their apartments. It was fine for me since I went to work every day, but WGM had converted our apartment's living room into an office, so Mary either had to be relegated to the bedroom all day or wander the streets of Anchorage during work hours. She was not happy. After a couple of weeks, the job ended and WGM gave me enough money for a plane ticket back to Colorado. Mary and I used the money to fly to Hawaii, where we backpacked around the islands for a couple of weeks before catching a flight back to the mainland.

After working for WGM, Mary and I moved to a friend's ranch near Laramie, Wyoming. I had met the rancher and his wife while working on my Master's thesis and they offered to help us out if we would look after their place when they were traveling. It was an idyllic situation, especially for newlyweds with limited resources. While taking care of the ranch, I pursued any geology-related employment opportunities I could find. However, permanent employment still eluded me and I had to settle for another seasonal position working in Arizona. While traveling to this new seasonal job I received a call from the Bureau of Land Management (BLM) office in Tok, Alaska. The manager of the office offered me a full-time job overseeing the BLM minerals program for the Fortymile Resource Area. Since full-time employment trumps part-time, I called the company that had offered me part-time work and told them sorry I wouldn't be showing up in Arizona because I was off to Alaska. Being geologists, I think they understood my decision.

In 1978, BLM was organized in districts and field areas. It had two districts: Anchorage and Fairbanks. The Anchorage district managed BLM lands in southern Alaska and the Fairbanks district managed lands in northern Alaska. The districts were subdivided into areas. Some of the area offices were co-located with the district offices and some had remote locations. I was in one of the remote locations, Tok, home of the Fortymile Resource Area. BLM, like any government entity, has reorganized numerous times since the late 1900s, so there are only a few of us old hands who might remember the time when there was a BLM office in Tok.

Since BLM wanted me to report for work ASAP (as soon as possible), Mary and I only had time to pack our essentials and call a moving company before I flew off to Alaska. Mary was tasked with driving our Volkswagen Beetle to Seattle, catching the Alaska ferry in Seattle, and driving from Haines to Tok a month later. I thought a VW Beetle with an air-cooled engine would be ideal for cold weather since it served us fine in Colorado. However, during the first winter I found out since the winter air in Tok was a lot colder than Colorado winter air, it didn't heat up much when it passed through the engine and into the cab of the car. This caused us to drive with one hand on the steering wheel and another scraping the ice off the inside of the windshield. Needless to say, we got a replacement vehicle the first chance we got.

Mary and I didn't exactly live in Tok. We lived at a

pump station about 10 miles out of town. The pump station was built by the military to pump oil from Haines, AK up to Fairbanks. At the time we were there, oil was no longer being pumped, but the station was habitable. It had two sets of apartment buildings, one with wood siding and one with aluminum siding. It was rumored the aluminum-sided buildings were sitting in Whittier en route to Guam when World War II ended. Someone decided there would be better uses for those buildings in Alaska. Whatever the true story was, the wood-sided buildings were a lot warmer in the winter than the aluminum-sided buildings. Even then, when temperatures in the winter would plummet to over 70 degrees below zero, our doors would freeze shut and people who had gotten out of their apartments would go around freeing the other residents by kicking at their doors to break the ice. The pump station was dismantled in the late 1980s or early 1990s. One of my neighbors in Anchorage was part of the crew that dismantled the station. Small world.

I was part of the build-up of government employees that occurred under President Carter. Almost everyone at the office was in their mid- to late- 20s and new to Alaska. Tok at the time had a population of about 700 people, scattered throughout the woods around town. It had one grocery store, a couple of hotels, bars and gas stations. Tok is located where one highway, the Glenn Highway, went west to Anchorage and one, the Alaska Highway, went north to Fairbanks. It was a 4-hour drive to Fairbanks and a 6-hour drive to Anchorage. We weren't exactly in the middle of nowhere, but it seemed like you could see it from there. The only channel on television was one devoted to educational television paid for by the State of Alaska Department of Education. Therefore, very few people had a television in our compound. Besides going to work, there wasn't much to do at night except read, listen to records and visit with the same people with whom I worked all day. It was a very quiet life, but OK. During the summers, I was outside as much as possible and in the winter, I was inside processing paperwork.

Winters were the time I would get all kinds of crazy phone calls. It seemed like people would hang around the bars in town and get these great ideas, then give me a call. One idea was from a couple of guys who wanted to have Jumbo jets full of Asian tourists fly into Tok. They were going to pick up the tourists and take them out into the bush so they could mine gold. I told them to go for it. Like most great ideas hatched while being lubricated with alcohol, nothing ever happened.

BLM is all about regulations and writing reports. The first job I was assigned was to write a report on a recreation site. My boss handed me the Code of Federal Regulations handbook and told me to get 'er done. It was very depressing. I had given up a seasonal job doing exploration to take a full-time job doing geology, or so I thought. To have to read regulations and write about a recreation site was not my idea of what a geologist was supposed to be doing. And, I discovered the majority of my job consisted of sitting in the office processing paperwork. At the time, it was an unspoken BLM policy if an employee was out in the field they were playing, whereas if they were in the office they were working. Sitting in the office, in a remote location, in the dark and cold of the winter months and watching the geological skills, which I'd spent so much time and energy trying to cultivate, slowly deteriorate was not my idea of how I envisioned my life. After two years I needed something different.

My impression of bureaucratic geologists through the Tok experience is these geologists have those jobs for certain reasons. Some geologists love office work, some don't feel comfortable in the field, and others want a home life. Some of the geologists working for the government have been laid off a number of times by industry and are looking for security. When metal prices are down, the government picks up a lot of really good geologists. Some of these former private industry geologists stay, while others go back to private industry after the prices rebound. As for me, in my late 20s, I wanted out of Tok and had two options. I applied to the U. S. Bureau of Mines (BOM) in Anchorage and with the BLM in Salmon, Idaho. It was a turning point in my life, I was either going to be a field geologist or a paper pusher.

After applying for both jobs, Don Blasko with the BOM called and asked if I was going to be in Anchorage anytime soon. I said, "Yes, I have some work to do in Anchorage." We arranged a meeting at the Federal Center and during the interview, he told me unfortunately, they didn't have any jobs available. I thought it was kind of weird he wanted to meet me, but now I guess he was just hedging his bet and wanted to know if all my limbs were intact. After the Anchorage meeting, I resigned myself to paper pushing and set my sights on Idaho. Fortunately for me, a couple of weeks later, Don called and offered me a job with the BOM, which I accepted

Mary and I moved to Anchorage in June 1980. My first job with the BOM was to assist Robert (Bob) Hoekzema on something called the RARE II project. RARE stands for Roadless Area Review and Evaluation. RARE was a study completed in 1972, in which the U.S. Forest Service evaluated areas for possible inclusion in the National Wilderness Preservation System. In 1977, a court ruled the study had to be redone. In 1980, the BOM and USGS received funding to evaluate the mineral resources of the Chugach National Forest as part of the second RARE study (i.e., RARE II). The evaluation was to be completed in four years. Because of the large size of the Forest, the BOM divided the forest into three areas and hired a field crew for each area. Each crew had permanent employees and seasonal employees. Bob's crew was unique because all of the employees were permanent and everyone except me was a lifelong Alaskan. Getting this job with BOM was finally the achievement of a dream, which was a full-time job where fieldwork was the main focus, hallelujah.

After working for Bob for a year, I took over one of the field crews. My idea of a great job did not seem to be the same as the people who worked on that crew. The crew boss quit because he wanted to return to private industry. He was an oil geologist, who had been laid off during one of the troughs. I guess he couldn't put up with government bureaucracy. His assistant also quit to pursue a doctorate at the University of Alaska, Fairbanks. To me, being a field crew boss was the penultimate job, so I gladly took on the responsibility and promptly hired Mark Meyer to help me. Mark was a geologist from Wisconsin who was working for BLM in Anchorage, so all he had to do was change where he reported for work in the morning.

Fieldwork for the BOM, which I did for the next 14 summers, consisted of evaluating areas of land for its mineral potential and determining the location, extent and economics of valuable minerals. Most of this work was done in cooperation with either the Alaska Division of Geological and Geophysical Surveys (ADGGS) or the USGS. The Surveys would map the overall geology, conduct the basic exploration work (i.e., stream and pan sediment sampling) and do geophysics. The BOM would follow-up on the sampling and map and sample any mineral deposits within an area. I was involved with the RARE II evaluation of the Chugach National Forest, the evaluation of the Kantishna Hills Study Area in Denali National Park and Preserve, Goodnews Bay Mining District, White Mountains Study Area, Valdez Creek Mining District, Yentna Mining District, Porcupine Mining Area, Strategic and Critical Mineral studies, Port Valdez area, National Petroleum Reserve Alaska, Koyukuk Mining District, Fortymile Mining District and ended up conducting some abandoned mine studies along the Fortymile Wild and Scenic River corridor and in northern Alaska.

I worked for the BOM from 1980 until 1995. I had a number of jobs from being a field hand to being a Section Supervisor in charge of field projects. Even when I was a supervisor, I always tried to get into the field as much as possible. I justified the fieldwork by saying I needed to know first-hand how the projects were progressing and field projects really needed extra help. My work in Alaska ended in 1995 when the BOM was being defunded by President Clinton. I was able to transfer to the BLM National Training Center in Phoenix, Arizona where I used all my Alaskan experiences to teach new geologists how to do fieldwork.

Finally, since 1995, I have been fortunate to return to Alaska many times for work and to visit the friends I made while living in the "Far North". Every time my wife and I return, we feel like we are going home. I think this attitude is based on the people we met and worked with, and my experiences working in the field in Alaska.

Loading a Skyvan

Chapter 4

Prefield Work

I was exiting the highway in Anchorage in order to go to the Costco when a guy in a beat up old car with visqueen for windows cut me off. We both stopped at the light and I got out of my truck and started marching up to that rude person. While approaching the vehicle I noticed the car looked like it was held together with duct tape and bailing wire and the person behind the wheel had long unkempt hair, a scraggly beard and was sitting in a pile of trash. I decided this wasn't the kind of person who would enjoy a stern talking to, so I did a U-turn and returned to my vehicle. The thing which made this incident so memorable was the fact I got so upset by a minor incident, which was abnormal behavior on my part. But, this abnormal behavior was a manifestation of the kind of stress I was under when I was trying to get everything ready for the field.

When I reported for work for WGM, much of the prefieldwork had been done, except for gathering and shipping field gear. For the rest of my career, it was up to me to be ready at the beginning of field season. In Alaska, there is a lot of planning and purchasing to accomplish before going to the field.

Once I knew how much money I had to spend on a project, the first thing I did was think about who was going to work on the project. For big projects, seasonal field crews were usually hired in Spring if there wasn't enough in-house personnel available for the job. This meant a trip to the University of Alaska in Fairbanks, which has the premier geology, mining and engineering programs in the State. Seasonal employees were usually college students or recent college graduates. After they were hired, they would report for work right after classes ended. This gave them enough time to take 2 weeks of safety training before heading into the field. For the BOM, they had to find their own housing, so most of the seasonal employees had friends and/or relatives living in Anchorage. I, of course, would try to hire the most qualified people. It was always nice when someone with a college degree and a couple of years of experience would walk through the door and would accept a summer job because I wouldn't feel like I had to do everything myself when I was in the field. But, many times the people hired would be just like I was when I reported to my first seasonal job, which was a strong back and with a willingness to learn. I had a saying, "A good grunt is one who thinks like the boss, but works like a superhuman." I always tried to find those kinds of people to work for me.

Prior to field season, transportation had to be contracted. If I could use 4-wheelers or boats, then transportation procurement only consisted of making sure those items were tuned up and ready to go. If fixed-wing airplanes or helicopters needed to be contracted, then costs for those services had to be factored into the project. Aviation contracting was a lengthy process and needed to be started as early as possible because the best aircraft were in high demand in the summer. Many times I didn't know the true cost impact on my project until all of the contracts were signed.

Once transportation was taken care of, I had to deal with field camps. Camp, office and field gear had to be purchased or gathered. Food had to be procured. Communications equipment, like radios, satellite phones and repeaters had to be purchased or contracted. Fuel supplies and delivery of fuel to the field site had to be arranged. I can't remember the true costs of field projects, but some of the BOM studies were multi-million dollar projects. I figured one time in the mid-1980s that each sample collected cost about $500.

Once everything was procured, all of the equipment would have to be packed up. This was Nathan Rathbun's role. He started out as a field assistant but moved into a warehouseman role as the BOM projects became more complicated. He would make sure all of the equipment was available and ready. He also helped the field crews pack up. He was in charge of 4-wheelers, a riverboat, rafts, pumps, sluice boxes, trailers, office supplies, kitchen appliances, radios, underground lights and self-rescuers. He also worked in the field and processed samples on an as-needed basis.

Our field crews would go out to the warehouse, pack-up and get the equipment to where it had to go for transport. In one instance Mark Meyer and I were going to take two 4-wheelers into an area north of the Glenn Highway. We were loading the 4-wheelers onto a trailer when the trailer tipped backward onto my left big toe, crushing it. In my haste to get loaded and into the field, I neglected to follow the first rule of working in a warehouse, which was to wear steel-toed boots. My toe was a mess, but since the whole field project was planned for that week, I persevered. Mark and I drove out to the Glenn Highway, unloaded the 4-wheelers, packed our field gear onto the back of them and in a little trailer and drove into the field. My toe was throbbing the entire time, and after we set up camp for the night, I emptied my rubber boot of the blood that had accumulated throughout the day. By the end of our time in the field, my toe felt better, but I had to get my toenail surgically removed a couple of months later. Usually getting the equipment into the field was a lot less painful.

Once in the field, we were essentially on our own until the end of the field project. This is not totally true because if I had some way of communicating with the office, which was not always the case, the office support staff hopefully had my back. The support staff included supervisors, the administrative assistant, secretaries, the warehouseman and any other employee who might be around the office. At one time or other, all of them helped out when there was a problem. One time I was at the Eagle airstrip field camp, which was in northwestern Alaska. I was playing darts with the seasonal field personnel. It was very challenging because the dart board was attached to a piece of plywood outside and we would stand 20 feet from the board. We would have to account for a 20 mile per hour cross-wind in order to at least hit the plywood. In the middle of the game, we heard a thump from the tent where the helicopter mechanic was staying. We rushed into the tent and found the mechanic on the floor. He was having a diabetic reaction due to low blood sugar. We got some juice and had him drink it and he was OK. After treating him and finding out this wasn't the first time he had experienced insulin shock (hypoglycemia) while in our field camp, I made a phone call to my office. I was concerned because the only medical training we had in camp was 5 geologists with basic first aid/CPR training. I asked the office to contact the Office of Aircraft Services (OAS) and get a replacement mechanic. I felt really bad because OAS would not replace the mechanic until I specified the reason for a replacement. I told them the reason and the mechanic was replaced. I never found out what happened to the mechanic, but hopefully, his company found him a job in a city.

WGM Camp

Chapter 5

Field Camps

"Every day a picnic, every night a campout," my boss said as I headed out to the field. Being a geologist or really any kind of scientist involved with field studies means living in camps. I have lived in tents, cabins, hotels/motels and lodges. My first experience with field camps was tent camping with my parents and the Boy Scouts. When I went camping with my parents, they did everything and I hoped to learn things through osmosis. Whereas, the Boy Scouts taught me to put together my field gear (boots, packs, clothes, tents, tarps, survival gear), how to buy food for myself and others that would last the length of the camping trip, how to set up camp, build a fire and cook food. I was also taught to deal with the elements and wildlife. These lessons set me up for a career as a field geologist.

Sizes of camps depended on a project's scope and budget. Some projects encompassed a large area with multi-million dollar budgets. These projects had field crews from 7 to 30 people with correspondingly large camps. Other projects with small budgets, usually had a 1 to 2 person crew with small camps.

Types of camps in Alaska depended on how the project area was accessed, transportation in the field, the size of the field crew and the availability of cabins, lodges or hotels. For project areas where aircraft were needed to access a field camp, tent camps were usually constructed. Some of these camps were large and elaborate, while others were minimal and consisted of a single tent. The biggest of these field camps were set up north of the Brooks Range at Ivotuk airstrip and Eagle Creek for the BOM Colville Mining District project. We needed large aircraft to move the camp gear. Mark Meyer and Nathan Rathbun did all of the ordering and organizing. They bought WeatherPort Shelter systems, which were half round tents attached to wooden floors. The camp had sleeping tents with cots, a cook and an office tent. The camp also had a washer and dryer, 2 refrigerators, a freezer, a 500-gallon water tank and a propane-powered shower. Of course, there was a generator and a water pump. Every tent had a kerosene heater.

Not only did the field camps consist of tents, but there was all the ancillary camp gear. There were tables, chairs, cots, lights, a generator, radio, associated kitchen paraphernalia and a toilet seat. There was also a library with reference books, reports and maps. There were boxes of field supplies containing sample bags, rock hammers, flagging, pens, pencils, notebooks, and empty boxes. An important box contained recreation gear. Most of these kinds of field camps had a volleyball net and ball, dart board, board games and playing cards. Whatever equipment and supplies that fit were placed in "Action-Packers". These were the plastic boxes of choice because they could take a lot of abuse.

An important aspect of living in tent camps was repelling mosquitoes, black flies called white socks and no-see-ums. This was usually done by using PIC Mosquito Repellent Coils. Coils were in every tent. They would be lit and hung from the ceiling or placed on a table. The smoke from them repels bugs. The coils supposedly contain a natural insect repellent with no health impacts, except on rats. In the early evening, we would close all of the canvas flaps on the windows and door of the tents, light a PIC and wait for the bugs to bite the dust. At bedtime, we would quickly enter the hopefully bug-free tent, unzip the flaps and watch the bugs crawl on outside of the window screens begging for a bite of flesh and blood.

When the project area was accessible by road and a helicopter was assigned to the project, we would try to stay in lodges or hotels as close to the project area as possible. Over the 18 field seasons in Alaska, I was able to stay in lodges or hotels 6 times. Staying in commercial facilities was a lot easier than hauling all of the camping gear and setting up and supplying a tent camp. All we needed to bring was personal and field gear and material for the office. We could move a 4 or 5-person field camp with a couple of SUVs or pickup trucks.

The most memorable lodge, which the BOM by way of a contract helped build, was in Kantishna on Moose Creek. Kantishna is located on the west side of Denali National Park and is accessible from the Park Rd. The BOM was evaluating the area and contracted with the Moose Creek Lodge, which no longer exists. The lodge owners built a cook shack with a wooden floor, erected canvas tents, a shower room, outhouses, a laboratory for samples, an office, volleyball area, and a helicopter landing zone with a fuel bladder. The camp had electricity and even a television with a VCR for watching movies. A hot tub, which was a big plastic shipping container, was also constructed overlooking Moose Creek. The owners of the lodge erected the camp for the summer and operated it for years afterward as a tourist destination. Today, the tent camp is gone and has been replaced by wooden cabins.

Staying in commercial facilities was wonderful because we had access to running water, sometimes indoor plumbing, electric lights, and beds. It was very luxurious. However, some people's idea of luxury is different than a field geologist's idea of luxury. In 1994, Joe Kurtak, Mark Meyer and I were investigating the Koyukuk Mining District. We stayed at a "hotel" in Coldfoot. It was composed of ATCO trailers left over from the pipeline days. ATCO is a brand of trailer construction companies use as temporary housing for their employees. The insides can be configured in various ways and some of the "hotels" in Alaska are made from cobbling these trailers together. The "hotel" in Coldfoot had private rooms, with private bathrooms, but more importantly, it had a restaurant, bar and a deck overlooking the Brooks Range. Every evening after working in the field we would return to Coldfoot, clean up, get a burger and a beer or something similar and sit on the deck watching the sun never set. We thought we were in the lap of luxury. However, I guess our standards were different from some of the tourists staying in the same place. We heard a lot of complaints about the facility. Instead of complaining, they should have felt privileged to have a roof over their head and someone to cook for them in essentially the middle of nowhere.

For projects that were accessible by road and a helicopter wasn't assigned to the project, we either set-up tent camps or tried to stay in a cabin. Many times the type of camp depended on how we were going to access the project area. When we worked out of boats or rafts, then camp was set up daily. We would bring a tent or tents, coolers, stove, cook gear and sleeping bags. Other times when we could access the project area on foot or with 4-wheel all-terrain vehicles (ATVs), we either had a base tent camp or sometimes were able to ask a local cabin owner for their permission to use their cabin. When I worked for BLM, I was able to stay at the BLM Fire Stations at Chicken and Eagle. They usually had extra rooms and as long as I brought some extra food and drink to share with the person in charge of the station, everything was copacetic. During one project, I got permission to stay at a miner's cabin north of the Glenn Highway. I found out while talking with him that he owned property three houses down from my brother's house in California. Small world.

When helicopters were used for a field project, we had to set aside an area as the landing zone or LZ. The LZ would ideally be flat, free of bushes and trees for a good radius around the helicopter. The LZ would hopefully be located towards the primary direction in which it would fly every day and within an easy walking distance from the camp. No one wanted the helicopter flying over camp and they didn't want the LZ so close to camp the rotor wash would knock the tents or anything that wasn't nailed down around. The helicopter in the morning was always taking off heavy because it had a full complement of fuel and personnel loaded down with breakfast. So, the pilots liked to have a long flat area to fly over when they were taking off. The LZ also had to have enough room for fuel storage. The ideal location for an LZ, of course, was an airstrip.

Fuel for whatever transportation we used was always problematic or kind of a pain. If a helicopter was used, then fuel for the helicopter depended on the location of the camp. Helicopters use either Jet A or Jet B fuel. Jet A is essentially a refined diesel and Jet B is a more volatile diesel mixture. Jet A works fine most of the time, but Jet B is the preferred fuel when the weather turns cold. Someone told me once some of the refiners in Alaska wanted to stop making Jet B, but since the military really likes Jet B for winter maneuvers, the refiners were encouraged by higher powers to continue to supply that fuel in Alaska.

Jet A or Jet B could be stored and delivered by various methods. The easiest method was to just buy fuel at an airport. I worked a couple of years out of Cordova and Valdez and it was nice not having to worry about procuring fuel. In remote areas, fuel was commonly delivered in 55-gallon barrels. The big balancing act in these remote camps was having enough fuel on hand to work and not too much so there were a lot of empty or full fuel barrels that needed to be flown out when the camp was moved. Another issue with fuel barrels was the environmental agencies wanted fuel barrels stored right side up and I wanted them stored on their sides. The problem associated with storing the barrels right side up was if it rained, the barrels might suck in water through their bunghole. A bunghole is located on the top of a barrel and is closed by a screw cap. Fuel is put into and taken out of the barrel through this hole. The possible problem with barrels laid on their side was the barrels might leak. Personally, and I can speak for the pilots also, I would rather deal with a leaky barrel than water in the fuel. I compromised and stored the barrels on their sides with the bungholes at the top.

If barrels were used, the helicopter pilot had to pump fuel out of the barrels using a hand pump. A person could endear themselves to the pilot by helping him pump fuel.

This was a hated job for everyone because it was hard on the arms and the pumper usually got fuel on themselves when they removed the pump from the barrel. So, after helping the pilot refuel the helicopter, not only would I smell of sweat, bug dope and dirt, but there would be the tell-tale Eau de diesel smell. The advantage to having barrels of fuel was they could be cached in remote locations, so the pilot wouldn't have to return to camp to refuel. This sped up work considerably and I always felt safer knowing the helicopter was nearby at all times.

For more elaborate camps, fuel was stored in collapsible 500 to 5,000-gallon bladders, which were placed on a fuel containment liner. The fuel was pumped using a diesel generator. Using bladders was a lot more efficient than dealing with drums, but the camp had to be in one place for a long period of time to justify the construction of the fuel site.

Fuel for boats, 4-wheelers and other vehicles was usually gasoline and was generally stored in 5-gallon fuel cans. When using these vehicles, we had to plan accordingly because once we ran out of fuel for this equipment, the project was over unless we were close enough to a town to be able to make a fuel run.

Moving a camp into the field, sometimes between camp locations and then back home again was always an interesting experience. For those camps with road access, the moves were relatively easy. We would load up the equipment in vehicles, drive to the campsite or commercial establishment and unload the gear. Sometimes, if the move didn't take too long, we could get some field work accomplished before dinner time. Going back to Anchorage was just as easy. We would pack up the gear, load the vehicles and drive home.

For more remote camps, moving to the field was more

complicated. If the campsite was on or near an airstrip, we would take the gear out to the Anchorage airport and it would be loaded into a cargo plane. A Short SC.7 Skyvan or a de Havilland DCH-6 Twin Otter was the aircraft of choice. Each would hold from 2.5 to 3 tons of cargo and were capable of landing on fairly short runways. I preferred using the Skyvan because the back opened as a ramp, so it was easy to load and unload. This was especially true when ATVs were used for field projects. Many times we just got what was available from the air charter company. For remote camps where there were no airstrips, camps were usually sited along lakes or rivers where an airplane on floats could land. If we were lucky we could get a de Havilland DCH-3 Single Otter on floats. It could haul about half the load of a Twin Otter but could land in very remote areas. If we weren't so lucky, we would get whatever aircraft the charter company had available on floats.

The field crew would access these remote camp locations via passenger planes. We would try to time it so we would arrive before the cargo plane touched down. When the cargo plane arrived, it was "Go" time. We would hurriedly unload the plane and start getting the cook tent erected. Getting a place where meals could be prepared was very important. It provided shelter and a place where the crew could rest and recuperate. The next order of business was erecting personal tents, then the office tent, then an outhouse. Someone usually got the short straw and had the wonderful assignment of digging the outhouse hole, erecting a shelter and installing the toilet seat. They always tried to site it downwind of camp and hopefully with a good view. It was always important to have a signal to show when the facility was occupied. When all the tents were set up, the ancillary gear could be moved into the appropriate accommodations. Luckily, we were working in the land of the midnight sun, so quitting time was sundown. We were extremely motivated to get the camp functional in the shortest amount of time because we knew that we couldn't start doing fieldwork until the camp was up and running.

Sometimes camps were moved to different locations throughout the summer. With WGM we moved to Tanana, which is at the junction of the Yukon and Tanana Rivers, then to Kokrines on the Yukon River, then to Sithylemenkat Lake, then to Lake Minchumina, then back to Anchorage.

An unforgettable move was when I was working for WGM. We moved camp in a Grumman G-44 Widegon from Sithylemenkat Lake to Lake Minchumina, which is about 160 miles southwest of Sithylemenkat. The Widgeon is a small amphibious airplane that can haul less than 1,000 pounds. An amphibious plane can either land on water or can deploy its wheels and set down on land. We needed an amphibious plane because we were going from a lake to an airstrip. It was late-August, which meant it got dark, so time was precious. We used the helicopter to move another person and me with our personal gear to Minchumina, which left the Widgeon to move the remaining four people and the camp. I got to take the helicopter because I was tasked with unloading the gear and setting up camp. The remaining four people packed up the gear and flew down in the last flights. Because a Widgeon can't carry much weight, and an ideal round trip took 3 hours, it was touch and go if we were going to be able to get everything moved in one day. The second to last flight came in with 2 people and some gear. They reported they didn't know whether or not the 2 people remaining at Sithylemenkat would make it with the rest of the gear on the last flight before darkness stopped all activity. It looked like their prediction was true when the Widgeon finally landed on the Lake Minchumina airstrip. All we could see was a plane stuffed to the ceiling with boxes and the pilot. When we unloaded boxes we found to our relief our two coworkers safely strapped to their seats. They said they were not going to spend the night camped out without food or shelter, so they willingly crammed into the cabin and let the pilot put the rest of the camp gear on and around them. A disaster averted by necessity.

As I said, moves were always interesting, especially when we moved back home to Anchorage. The day of the move had to be scheduled with the charter air service months in advance. I always tried to reconfirm the scheduled service prior to the move. Days prior to the move we would have feasts because we didn't want to ship food home. We would also start packing everything that wasn't immediately needed to do fieldwork or sustain the camp. As we got stuff, like extra field supplies and the library, packed we would place it in an out-of-the-way area. The day before the move, we would pack the office tent and anything that wasn't absolutely needed for that night. On the morning of the move, we would pack the cook and personal tents and our own gear. As we were packing, we had our ears attuned to any buzzing that might signal the approach of an airplane. It was always a joy and relief when the scheduled plane arrived at the campsite.

Waiting for planes to get you to and from places was always stressful because I have been stranded in Alaska for days waiting for the weather to break. Luckily the results of each instance have been minor. One of the places I was stuck was at the Red Dog Mine. Joe Kurtak, Mark

Meyer and I had gone out to look at the mine. After we had been shown everything at the mine and mill site, we sat and waited for our transportation home. Unfortunately, no planes were able to fly between Kotzebue and the mine site for a couple of days. Fortunately, it was sunny at the mine site and the only thing we had to overcome was frustration and boredom. Regrettably, because I was stuck at the mine site, I missed wishing my father Happy Birthday. He died before his next birthday.

A final note on field camps is the most opulent field camps I visited were those associated with big mineral operations, like at the Red Dog Mine and Prudhoe Bay oil field. I call them field camps because the buildings at both locations are substantial but temporary. They are built to be disassembled at the end of operations. Also, the personnel working at the camps are only there for short periods of time. There are various work schedules, but one of the most common was two weeks on and one week off. Everybody who works at these camps has a home somewhere else. Some people have homes in the lower 48 and commute, weather permitting. Even though the camps are elaborate and I was jealous of their day-long food service, indoor plumbing and formal entertainment, they still contain all of the components of a field camp. Despite the amenities, I really enjoyed visiting these fancy camps.

Geologist eating lunch at Goodnews Bay

Chapter 6

Food

"Coffee," I moaned as I dragged my weary body out of my sleeping bag, got dressed, staggered over to the cook stove, filled the coffee pot with water, loaded the pot full of fresh coffee grounds and turned on the stove. It was the 20th day in a row of working at Goodnews Bay and my body was dragging, but I knew a good dose of that black caffeinated liquid would help me face another 12-hour work day. Coffee is the liquid that fuels field camps. Many geologists are powered by caffeine. So, large amounts of coffee, especially in the morning always have to be available. After enough coffee was in me, I could start getting breakfast ready for the rest of the crew.

As I stood cracking two dozen eggs into a large skillet where onions and green peppers were sizzling, I kept thinking even though I could make palatable food, the best food I've had in field camps was prepared by a hired cook. The cook would prepare a hot breakfast that might include eggs, bacon, pancakes, cereals or anything the cook might dream up. After breakfast, I would make up my lunch, which usually consisted of a sandwich, candy bars, fruit and something to drink. The cook would always have a hot meal waiting for us when we got back from work. The cooks were usually creative and the food was plentiful and good. I always thought the cook in a field camp has one of the hardest jobs. They had to get up hours before the rest of the crew in order to have breakfast ready and put out the lunch food, wash up after breakfast, put away the food and then start planning for dinner. After dinner, they had to wash up and plan for breakfast. They usually never got a day off because without them the crew wouldn't eat. The cook was also a person to whom everyone was nice because life was not good if they were unhappy.

All of the cooks I worked with were wonderful people, but the cook in my first camp in Alaska was the most unforgettable. Not only was she a graduate of a culinary school but she was also a singer/songwriter. Therefore, we had entertainment when she felt like performing. She had come to Alaska and stayed because there was a surplus of men. Her saying was: "The odds are good, but the goods are odd." I guess the goods weren't odd enough to leave. Not all cooks in later camps were as unique and multi-talented as this first cook

As the eggs began to cook, I thought about the second best food I had when camping out. It was food I or my crew prepared. When I worked for BLM, I was able to go to the firefighter mess hall and get anything I wanted to eat. I would pack the food into coolers and boxes for field projects that lasted overnight to a week long. My boss did complain once when I wanted two pork chops for dinner. He didn't think a 20-something-year-old could eat so much. I proved him wrong. For other field projects, I would go to Costco in Anchorage and buy enough food for the extent of the field project.

Breakfast was usually bacon and eggs. Lunches for me were usually sardines, which came in varieties (mustard, oil, ketchup, hot sauce) and pilot bread, which is a round hard cracker, candy bars and fruit. Dinners were some kind of meat, fresh vegetables, however long they lasted, and some kind of dessert.

The thoughts about food continued as the rest of the crew got their plates and I dished out the eggs. I decided the least best food was that which was procured in hotels and lodges. A person would think the commercial establishments would be the best, but I had some "memorable" experiences associated with hotels and lodges. One day I came out to work on the Valdez Creek project. The field crew was staying at a lodge on the Denali Highway. I got the lunch packed by the lodge. A crew member said they had a pork roast for dinner the previous night. So, I was looking forward to a good pork sandwich. Well, at lunchtime, I took a bite of the sandwich and it was kind of mushy. I spit the bite out and took a look between the slices of bread and found a big slice of pork fat. Needless to say, the sandwich was sent to feed any critter less picky than me.

Another time I was staying at a lodge on the Steese Highway. One morning the crew got up for breakfast and the lodge owner was nowhere to be seen. I had to wake her up and tell her my crew needed breakfast so we could get to work. She went back to the kitchen and rattled around and brought out food that looked like it was reheated from the previous night's dinner. We ate it, with no ill effects, but it wasn't a very pleasant experience thinking the food had probably been sitting out all night. Lodge and hotel ownership is hard in Alaska and there is a high turnover rate. So, I'm sure the places I mentioned have had a lot of new owners since I stayed there in the 1980s.

One of the most memorable dining experiences was when I was working out of Glacier Island in Prince William Sound. The crew was staying in tents and eating at a mess hall on the island. The mess hall was set up to feed tourists who came out for the day to experience Columbia Glacier. Staying in a tent was OK, but eating in the mess hall was challenging, to say the least. The operator served the same food every day because tourists only ate there once. The operator served baked salmon, baked halibut, coleslaw, potatoes and corn for dinner. For the first couple of nights eating salmon or halibut, we thought, "This is great." But after a week of the same food and the fact that when we got back from the field there wasn't a green vegetable left, the greatness wore off. We started working as late as possible and joking among ourselves, "I wonder what we'll eat tonight?" Luckily the project only lasted 2 weeks.

One issue when I worked for the government was I only got so much money for food. The money was usually enough to survive on, but I couldn't eat too extravagantly if I didn't want to eat into my pay. So, I usually ate in cafes when living in hotels. One cafe in Cordova, which had good food had a nightly special, which was usually really good, which meant generous portions at a reasonable price. The only problem was our crew usually worked so late that by the time we got to the cafe, the special was gone; therefore, we were relegated to ordering off the menu. The menu items were usually not as noteworthy as the special.

I had a funny experience while staying at the Sunrise Inn on Kenai Lake. The BOM had a contract for food for the first couple of weeks we stayed at the inn. During that time we ate like there was no tomorrow. Breakfasts would be omelets, orange juice and toast. Dinners would be salads, steaks and dessert. Once the contract was over, the owner of the Sunrise Inn was shocked when instead of wanting an elaborate breakfast, our crew ordered the cheapest things on the menu, which was usually a short stack of pancakes. Also, the shock continued when for dinner, we ordered a hamburger with water instead of steak and a soda. The whole crew was supporting households back in town, so no one wanted to use their paycheck, which the family back home needed to live on, to buy elaborate meals.

Camp at Eagle Creek Airstrip

Chapter 7

Camp Life

"Who's up for a game of volleyball," shouted Tom. "I'm in, once I finish my daily fieldwork report and figure out what we're going to do tomorrow," I shouted. Thus started another night in our Yukon River field camp. Living in a field camp necessitates that people from various backgrounds and who are mostly strangers have to live, work and play together for months at a time. The routine is much the same in all field camps. Out of the sleeping bag between 6 and 7 a.m., get dressed in field pants, shirt and boots, breakfast, then meet at the office tent for the day's assignment(s). Back to the tent for field vest, pack and field belt. To the cook tent to make lunch and then off to work via a helicopter, ATV, truck or on foot.

Most large field camps I worked in had five geoscientists or engineers. Usually, there were one or two permanent employees and the rest seasonal employees. The project leader and the project geologist were usually the permanent employees. The project leader ran the camp. It was an unenviable job. They were in charge of personnel management, which not only included making sure everyone got paid but that everyone got along. They also had to track the budget; ensure there were enough food and supplies available and if not order more; make sure all of the equipment was working; oversee the quantity and quality of the field and office work; deal with anything the people back in the office might demand; and if that wasn't enough, the project leader had to go out and do fieldwork. Everything associated with the camp was their responsibility.

The project leader's right-hand person was the project geologist. The project geologist's main duty was to ensure the work outlined by the project leader was being accomplished. However, the project leader also hoped that the project geologist would point out any problems or deficiencies in the camp or between personnel. A good project geologist could relieve the project leader of some of the personal problems associated with people living in camps while the leader dealt with all of the administrative stuff.

The seasonal employees, field hands or grunts had the best jobs. They just had to go out and do their job and at the end of the workday, play. When I was a grunt I always wanted to be the boss, but after being a boss, I realized that being a grunt was a pretty darn good life.

I enjoyed working with teams of five because only four people would usually go into the field on a daily basis. This meant one person could always remain in camp. The person left in camp could sleep in, take a bath or shower, do laundry and just hang out. Bathing oneself in remote camps was always an experience. Many field crews set up a shower tent or tarp on a frame. A person would heat water on the stove and put it into a bladder attached to a showerhead. Other times, if camped next to a lake, the lake became a big bath. Interior Alaska is usually warm, in the 80s and 90s during the summer, so swimming in a lake at the end of the day was wonderful. However, many times bathing was just heating water on a stove, pouring it into a 5-gallon plastic bucket, then using some kind of ladle to get wet, then soap up and rinse off. The most elaborate camps had on-demand propane water heaters. How decadent can one get?

Laundry, except when living in a lodge/hotel or elaborate camps with washers and dryers, was accomplished by dumping clothes in a 5-gallon bucket, adding soap and water and agitating it with an inverted cone attached to a stick. The clothes would be rinsed in clean water, wrung out by hand and hung on a line to dry. Laundry could only really be accomplished when there was a day off and the weather cooperated.

A day off was really never a day off because everyone was expected to make sure all of the office work was done, camp was cleaned up and engines fueled; samples boxed and ready to go if and when a supply plane came in; help the cook around the kitchen, if he or she would allow it; and tote supplies upon arrival of supply planes. An incident that always stuck with me was the time we were getting barrels of jet fuel flown into Sithylemenkat Lake from Fairbanks. The lake is located about 135 miles northwest of Fairbanks and would normally represent a flight of a little over an hour. We had run out of jet fuel, so everyone was in camp when a plane flew in with the fuel. It was an old plane on floats with four barrels of fuel. The pilot and his friend didn't arrive until late in the afternoon because the pilot said he got lost trying to find the lake. I guess the plane was old and the only compass they had was a hand compass that didn't work because of the plane's massive iron engine. They also did not bring anything to unload the fuel barrels. Luckily we had enough people to manhandle the barrels out of the plane, which sat about four feet above water level. After we unloaded the fuel barrels the pilot took off. He returned the next day with a new load of fuel and I was in camp on my day off. Because he left his friend at home and all of the field crews were out working, it was just me and the cook in camp when he motored onto the beach. I asked him how he was going to get the barrels out of the plane. The pilot said he didn't think he needed to bring ramps today because it hadn't been a problem the first time. Needless to say, it was a problem getting fuel barrels weighing about 315 pounds out of a plane sitting four feet off the water. The pilot, cook and I rigged a ramp using scraggly black spruce trees we cut down and with a little cussing, we were able to roll the barrels onto the shore.

Camp came alive when the crews started returning from the field. Everyone would come back, drop their packs at their tents and take the samples from their packs to the office tent. Depending on when a person got back, they would either do the office work immediately or wait until after dinner. But, one of the first things everyone always did was take off their boots and change their socks. Because Alaska has a lot of streams, bogs and generally just wet ground, most people who work in the field wear "Alaskan tennis shoes", which is the name for rubber boots. Extratuffs were the boot of choice. Very rarely did I wear my leather hiking boots because if my hiking boots got wet, which they invariably did, it was really hard to dry them out while living in a tent. And, if the boots didn't dry out, they would essentially rot and fall apart. Most geologists spent too much money on a good pair of boots to see them trashed. So, almost everyone wore rubber boots, which invariably led to sweaty feet. Getting those boots and damp socks off their feet was, therefore, a priority. Many geologists I know got foot fungus from wearing rubber boots all of the time. This was very common for people working in Southeast Alaska. Luckily, I never got it.

Dinner was served in the mess tent if living in a tent camp with a cook. It was family style with whatever the cook decided to serve. Sometimes menus were challenging for the cook because it seemed like in the late 1970s people started having dietary issues. So, some cooks had to contend with crews with vegetarians, vegans or food allergies. Usually, the cooks were very accommodating. Personally, I was just glad I was not cooking and would eat anything put in front of me. I ate bear heart and liver, lynx, rabbit, ptarmigan, overdone and underdone food. If there wasn't a cook, then the designated cook for the night would start preparing dinner for the rest of the crew soon after everyone returned. We'd all pitch in to help. In these instances, nobody ever complained about the food because they knew whoever complained would become the cook until the next person complained. When living in commercial establishments, we usually would just drop our field gear, change clothes and get to the eating establishments before they closed for the night.

After dinner, the crew would either hang out around the cook tent or write up the results of the fieldwork at the office. All samples had to be plotted and a daily report written. The daily report writing exercise gave me a chance to summarize what I had discovered and think about the implications, if any, of what was accomplished. Any maps drafted in the field had to be redrafted in ink. The daily reports and redrafted maps were put into notebooks and filed chronologically. These daily reports were used to eventually write up the yearly report. It was therefore very important for seasonal employees to be as comprehensive as possible because they probably wouldn't be around to ask about their results after the season was over.

While the seasonal employees usually went off to recreate after the office work was done, the project leader and project geologist went over the work that was accomplished and decided on the next day's work. After the next day's work was lined up, the project leader and project geologist could relax. Because field crews were mostly composed of men and women in their 20s and 30s there was usually a lot of energy to be expended. This sometimes took the form of drinking. Geologists are notorious for the amount of alcohol they can consume. Working for private industry, the company would usually supply the camps with beer and sometimes hard liquor. However, working for the U.S. government, every individual would be responsible for their own alcohol, which usually took the form of something with a high alcohol content and would fit in a compact container. The government-contracted supply planes did not bring alcohol, so whatever someone brought to camp at the beginning of the season might have to last 2 months. Therefore there was usually not too much hard drinking in camp. Other recreation included volleyball. A game was usually an easy thing to organize. There were always card and board games to be played. Many times the games played were dependent on the weather or how hard people had worked.

A highlight of camping was when the supply plane arrived. Not only would it be carrying needed supplies, but also mail. I tried to write something to my wife every night and would gather all of my musings into an envelope the night before a supply plane was scheduled to arrive. In the morning, everyone would leave their letters with the person remaining in camp for the day. This person would ensure that those letters got onto the plane or there would be dire consequences for that person. Upon arriving back at camp, everyone would rush to the cook tent to see if they received any letters or packages. What joy it was to get a letter from home. The letters also gave me something to comment on in my nightly ruminations besides talking about walking, sweating and swatting bugs.

The rowdiest camp I worked in was at Kantishna on the west side of Denali National Park. The geologic consultants the BOM hired to evaluate the Kantishna area, contracted with the Moose Creek Lodge to provide food and lodging for about 30 people. The people included geologists, geophysicists, drillers, various helpers and a helicopter pilot and his mechanic. Sometimes the camp had more than 30 people because whoever was mining in the area, liked to stop by for entertainment and visit the lodge owners. Some of these unofficial people were real characters who when mixed with the official people caused a lot of mayhem. Two of the miners living in a trailer near camp liked to shoot guns, so they organized a marksman contest near their trailer. The miners had found some dynamite and decided to have a contest. Everyone who wanted to compete would get to shoot at their own stick of dynamite. The contestants would start 20 yards away from the dynamite and get one shot. If they didn't hit the dynamite, they would have to walk up 5 yards and shoot again. Another miss, another 5 yards closer. If they were a real bad shot they could blast away from 5 yards until the dynamite exploded. I didn't participate but could hear the shots and explosions from camp. Another time, these miners tried to scare the people in camp by setting off a blast right next to camp. The only harm they did was blow out the windows of their own trailer. At the least, these characters broke up the monotony of camp life.

Until the advent of satellite telephones, which we had in camps in the 1990s, communications with the outside world was accomplished via the bush radio or single sideband. The bush radio was like a giant Alaska-wide party line. Nearly everyone in the bush listened to conversations between loved ones and between various bush camps, villages and towns. Everybody used the radio. One conversation I remember was between a bush camp and a town. The person said, "We need Preparation H. Yes, Preparation H, this is not a joke, this is an emergency." I could almost hear the chuckles throughout Alaska when the listeners heard this plea. If I needed to get in touch with someone, I could always get patched through to a telephone line. It was the old version of cell phones, but a lot more entertaining. The biggest problem was remembering to say, "Over" when I was done talking. Of course, if the field camp was located near a town, people would just use any available telephone.

Bedtime was usually between 10 and 11. Because I had been working hard all day and after playing a rousing game(s) of volleyball, sleep was usually not a problem, even if it never really got dark. Depending on where I was working, there might be anywhere from 8 hours of dusk in May to 4 hours of dusk in July. When working north of Fairbanks, the sun really never set. If it did set it was because there were mountains around the camps.

Human interactions are an important aspect of camp life. Working, eating, recreating and sleeping around 5 to 30 people for months at a time can be challenging. There were much laughter and good times, but there was also a share of hurt feelings generated usually due to callousness. Geologists, particularly field geologists pretty much fall into the same category. They are generally not people people unless they have been drinking. People who like to be around other people are not the types to go tromping through the woods for days at a time by themselves. They also have healthy egos. When a geologist is working, they have to make countless decisions. They have to be a person who is not afraid to make a decision and sometimes defend it before their peers. A geologist is also a skeptic, while at the same time being an optimist. When a geologist is looking at a mineral property, they have to question whether or not the property might be economic and ask themselves what needs to be done in order to prove if it is either a good or bad property. But, they also have to be optimistic enough to think of the possibility the next sample or next drill hole will make the property the greatest mineral discovery since the Comstock. So, a camp full of egotistical skeptic loners always made for some kind of drama occurring over the course of a field season.

Every now and then drama took the form of camp romances. Camp romances between geological personnel were usually not encouraged because of the possible distractions associated with the people in those romances. But, romantic occurrences were to be expected when in-shape 20-somethings were in close proximity. Add to the mix a swash-buckling helicopter pilot and romances will blossom. When I worked for WGM, they ended up redistributing field personnel mid-summer to put an end to camp romances. Camp romances between field personnel and the public were tolerated. When I worked for the BOM in Kantishna, our camp had camp followers. These were women who migrated to the camp from I don't know where and took up with some of the field

personnel, some of whom were married. High drama occurred when one geologist's wife, who was concerned
about her husband, suddenly showed up in camp unan-
nounced.

A helicopter pilot, who worked in Alaska during the
summer told me he was single by distance. The further
he got from his wife, the more single he became. Another single female geologist told me she always hooked
up with the helicopter pilot if given the chance because
every male geologist knew the helicopter pilot always
got the women, so there were no hard feelings in the
camp. Essentially she sacrificed herself for the good of
the camp. One of the pilots I had for the Prince William
Sound study was Ted. He seemed to have a girl in every
town. I never had to worry about Ted. He was pretty
self-sufficient when it came to entertainment.

The author working in the Brooks Range

Chapter 8

A Field Geologist Is...

Geologic fieldwork consists of more than just tramping around woods, deserts, mountains and valleys. I was always told, and it's probably more than just an urban myth, that geology is one of the most satisfying professions, but it is really hard on family life. Divorce is very common among geologists. Fortunately, I dodged that bullet, thanks to the forbearance of my wife.

The best field geologists are the ones who have seen the most geology. When I hired on with WGM, I looked at my boss and thought, "Boy, he knows everything because he's been a geologist for 6 years." Now, after 40+ years of being a geologist, I realize there is still so much I haven't seen and there is so much more to learn. Most field geologists go on field trips during conventions or go on vacations that pique their geological interests to increase their knowledge of the earth. Personally, I have vacationed in Hawaii in order to look at the volcanoes; Malaysia to look at the tin mines; Australia for their opal, sapphire and coal mines; hiked into Annapurna in Nepal and viewed the transition from sedimentary to metamorphic to granitic rocks; looked at the Greek temples composed of not only marble, but fossiliferous limestone; viewed the ancient rock quarries of Sweden that supplied much of the building stone throughout the Baltic; saw the gold, amber, malachite and azurite used to decorate the palaces of St. Petersburg; hiked into the Chilean Andes and viewed the alteration associated with the famous Chilean copper deposits; looked at the folding associated with the sedimentary rocks of the French Alps; and of course visited many of the geological wonders of the United States, like the granites and glaciation that compose Yosemite, the geysers and volcanic activity that are Yellowstone, the erosive forces of the Colorado River at the Grand Canyon and the Colorado Plateau country of Utah.

Why would someone become a field geologist? There seems to be a variety of reasons. Some people like rocks, others minerals, while others just like the out-of-doors. Some people get into geology because they are mountain or rock climbers. Others because they grew up around mining. I may have gotten into geology because it is in my DNA. My father and his brothers grew up in Idaho during the Great Depression. Their mother cooked in the mining camp of Atlanta, Idaho. My dad and his brothers went to work in the mines when they were teenagers. My dad would tell stories about working in an underground mine in Warren, Idaho. He said it was hard work punctuated by the excitement of scrambling out of the mine before every dynamite blast. His older brother, Ed, who worked in Nevada and finally in the California gold mines, didn't like his little brother working in those old mines so he got my dad a summer job working in the mill at a gold mine in the California foothills. Every summer after high school, my dad would go to work in the gold mill with Ed. My dad left the mines after graduating from

high school when he went to work as an American ser-
viceman for Uncle Sam in the Pacific. After World War
II, he stayed working for the US Air Force as a civil-
ian. Ed, who married a woman whose grandfather was
an original California 49'er left the mines to work in the
shipyards during the war. He kept his hand in mining until the 1970s when the Federal government took away his
gold claims, but that's another story.

Field geology is not for everyone. The characteristics of a field geologist include being introverted, confident, curious, analytical, independent, one who thinks
of themselves as a tough guy or gal, optimistic and one
who likes to search for buried treasure and solve mys-
teries. Geologists seem to be more nerdy than athletic
but in an outdoorsy way. Maybe it's because their field
vests resemble the pocket protectors worn by stereotyp-
ical nerds. Most field geologists are in good shape, at
least after a couple of weeks in the field. A field geologist doesn't like being in the field with someone who will
run rings around them, so they try to maintain some kind
of shape all year long.

I found I could never make the mistake of looking
at someone and deciding whether or not they could do
the work. I have worked with people with every kind of
body-type. I am tall, so I have certain advantages and
disadvantages when I'm in the field. One advantage is
I can quickly cover a lot of ground. This is really apparent in those areas of Alaska with sage tussocks. Sage
tussocks are plants that usually grow in swampy conditions. They are about one to two feet tall and one foot in
diameter with a grassy top. They grow independently of
each other with about a one-foot space in between each
tussock. If I got some momentum going, I could walk
on the top of these tussocks and cross a field of sage tus-

socks fairly rapidly. However, I worked with shorter people who would have to walk in the water and mud around the tussocks, which took a lot more time for them to cross the same field. Another advantage a tall person has is in crossing streams, especially small streams which I could walk or easily jump across.

There are distinct disadvantages to being tall. Ingress and egress from aircraft is a problem, especially if I had to fold myself into the back of a helicopter. Working in underground mines is hard on the neck because most abandoned mines are less than 7 feet tall. Many of the mines are around 6 feet tall, so I had to be careful to avoid slamming my head into the ceiling when I was underground. I attribute my current neck problems with how many times I wasn't so careful and would find myself on my butt on the mine floor after accidentally contacting the ceiling.

Being short also has its advantages and disadvantages. Getting in and out and riding in aircraft is easier for shorter people. Also, pilots like people of short stature because they were usually lighter. Since aircraft can only carry so much weight, the less people weigh, the more fuel and rocks the aircraft can carry. Short people also can investigate underground mines easily without worrying about hitting their heads. People of this stature were usually more nimble than us tall folk. The disadvantages are ones I already mentioned.

No matter what body type a field geologist possesses, many possess a tough guy or gal attitude. Someone who personified the tough guy aspect of a geologist is Dr. Tommy Thompson. When I was in graduate school at CSU, I talked to him about his summer fieldwork in New Mexico. He told me he miscalculated the time it would take to finish his fieldwork and darkness caught him far from his camp. So, I asked, "What did you do?" He said he made himself as comfortable as possible, ate some food, made a bed out of pine needles and spruce bows and went to sleep. I then asked, "You went back to camp in the morning, right?" He told me no, he got up when it got light, ate some food and did a full day of fieldwork before returning to camp for the night. I think this story epitomizes the "tough guy" geologist. I reconnected with Tommy when I moved to Nevada in 2004 and found out he's still a tough guy. Small world.

Women I worked with were also tough or they wouldn't be field geologists in Alaska. I worked with a female geologist who would come in from the field and go jogging. She also lost her rock hammer one day. Losing stuff like rock hammers, knives, and other accouterments is a common occurrence for a field geologist. After she lost her rock hammer, she just used harder rocks to continue her work breaking rocks. Another woman, a geologist from Fairbanks, interviewed for a seasonal job with the BOM. She didn't get the job because she showed up for the interview on 2 broken legs. When asked how she broke her legs, she said she was mountain climbing with her boyfriend and their anchors came loose and they fell. She broke both legs and her boyfriend was seriously injured. She crawled miles out to a road to get help. Unfortunately, help didn't arrive in time to save her boyfriend's life. I would have given her a job but didn't have enough money to hire someone who was not going to be able to walk. I worked with a woman, Denise Herzog, whose father owned a placer mine. He loved to tell people that his daughter was such a hard worker that when people saw his daughter at the mine, they mistook her for his son. He would correct them with a knowing smile.

What kind of degree must a person have to go prospecting or do fieldwork? A college degree really isn't necessary. As I said, the best person is the one who has seen the most geology. In Nevada, some companies hired prospectors to find gold deposits. One of the best gold finders I know is Nathan Rathbun. He started working for the Forest Service in Alaska right out of high school. The BOM hired him as a helper and eventually the warehouseman. He is a guy who works hard and has a nose for gold. He was always a great guy to have on a field crew because, besides his propensity for finding gold, he could troubleshoot any kind of mechanical equipment.

Although a degree isn't necessary to do fieldwork, in reality, in order to get a full-time geologist job, a college degree in geology or earth science is needed. A Bachelor's degree can pave the way for a geology job. Lately, a woman I know who has a degree from a college in Georgia was working as a waitress in Yellowstone when she waited on a geologist from Nevada. They started talking and she said she was looking for a geology job. He asked her for her contact information. A couple of months later, she got a call from the geologist who offered her a job at a gold mine in Nevada. It seemed a man they hired did not like the work and quit after a couple of weeks, so the company was desperately searching for someone who might be better suited to the job. Lucky for her she had met this geologist and was available. She's been working as a geologist ever since.

An advanced degree, like a Master's or Ph.D, can open more doors for a person. My experience is a Master's degree gives a person an advantage over a person with their Bachelor's degree when it comes to scientific methods, geological theories, field methods and report writing. A person with a Ph.D has advantages in getting jobs with universities, the Geological Surveys (state and federal), and some companies and consulting firms. It also does not matter if a geologist graduates from a prestigious school. I met a geologist with a Masters degree from John Hopkins who had never worked with a compass and showed up in the field wearing loafers on his feet. I guess he decided fieldwork was not for him and he became a regulator with the BLM. It does seem small schools whose professors actually teach students churn out more field geologists. An alarming trend lately is colleges and universities are eliminating their field camps and letting the students take computer mapping courses instead. A complaint I hear from mining companies is their recent hires want to sit in an office, working on their computers instead of being in the field.

One thing I've always tried to do was hire the smartest people I could find, either as field assistants or permanent employees. Some supervisors are intimidated by smart people, but I found if smart people are hired they made me look good by doing exceptional work. Luckily, I worked for the BOM at a time when most of its employees were top-notch.

The author with all of his field gear

Chapter 9

Field Gear

"Rats," I said, or maybe I used a stronger expletive. I was at my first prospect on the first day of the field season and didn't have a magic marker. Oh well, I'd make do and rectify the oversight when I got back to camp. Then I wondered what else I'd forgotten.

A field geologist is self-contained. They usually carry everything they need with them. The primary piece of field gear for any geologist is a vest. The vest has many pockets where one can carry pens, pencils (colored and mechanical), black markers, flagging, hand lens or lenses, clinometer, altimeter, magnet, measuring tape, field notebook, a notebook with geology and engineering information, scratchplate used to determine the crushed color of a mineral, mosquito repellent, rulers and an acid bottle. I usually carry my field maps in the back pocket of the vest, but I've also been known to load the back pocket with rock samples. The "cool" geologists like their field vests to be tan in color, but I always preferred wearing the brightest color possible, which was usually orange. I wanted to be seen by other people. If working in two's and trying to take bearings on the other person, it is really hard to see a person wearing tan against brown trees and sage colored bushes.

A geologist usually has a clipboard with a cover which is used to carry maps and drafting paper. It can also be used as a portable desk. Working in rainy conditions, I stopped carrying a clipboard and carried all of my maps in a plastic bag. The plastic bag slipped easily into a pocket in the back of my field vest and was waterproof. Field notebooks are composed of write-in-the-rain paper, so they don't need as much protection as paper maps.

A geologist also has a field belt on which they have their compass, usually a Brunton-type compass or the Brunton pocket transit. The compass or pocket transit was patented in 1894 by a Canadian-born geologist named David W. Brunton. It is manufactured by Brunton, Inc. of Riverton, Wyoming. A Brunton-type compass is a surveying tool that can not only be used to show the direction but can also show angles. It is so accurate if it can be steadied sufficiently that a person can read the direction exactly or at least within a 1/2 of a degree. I would also have my rock hammer in a holster, a knife and a radio on my field belt.

A geologist's rock hammer is one of their most precious and personalized possessions. I spent a lot of time finding the perfect rock hammer, which seems kind of silly to a non-geologist, but since there are so many kinds, it makes sense to geologists. A typical geologist rock hammer has a blunt (hammer) end and a pointed end. The pointed end can either be wedge-shaped or pointed. A hammer with a wedge-shape is usually carried by a soft-rock geologist and a hammer with a pointed end is carried by a hard-rock geologist. Soft rock geologists usually only study sedimentary rocks, while hard rock geologists study all kinds of rocks. These hammers usually weigh either 14 or 22 ounces, with the most common being the 22-ounce weight. If a chisel is going to be used, many geologists use a crack hammer, which is a 2 to 4-pound hammer with no pointed ends. Personally, I like to carry a 2-pound blacksmith hammer I buy at hardware stores. It looks like a crack hammer, but has one wedge-shaped end and is a lot less expensive than a "geology" hammer, which is important because I will invariably lose it. Why does a geologist hit rocks with a hammer? Because they can and it's what they do. The real reasons are a fresh surface is needed to be able to identify the minerals in a rock, there just might be some hidden treasure inside the rock and it's what they do.

An accessory to a rock hammer is some kind of eye protection. When a rock is hit with a hammer, chips inevitably fly into the face of the hitter. Even with eye protection, geologists quickly learn to turn their head away from the rock they are hitting unless they want to deal with a sliced up face, chipped teeth or broken eyewear. If I was chipping rocks with a hammer and chisel, there was no way to protect your face, so I just hoped the chips would fly somewhere else and wore eye protection in case they didn't. The agencies I worked for provided prescription safety glasses. They weren't stylish, but at least I didn't get my $400 pair of glasses trashed breaking rocks.

A backpack is a necessity for a geologist. In Alaska, people carried between 20 and 60 pounds in their packs. I would usually carry my rain gear, poly-pro jacket, first aid kit, radio with spare batteries, water bottles, lunch, tarp, extra clothes, sample bags, leather and/or cloth gloves, camera, a head net for bugs, chisels and ammunition. When I was working in areas with a lot of underground workings, I would also carry a hard hat, lights, spare batteries for the lights and a 100-foot long measuring tape. All of these items would be put into a big plastic bag in the pack. There also had to be enough room in the pack to put 10s of pounds of rock samples. One reason it pays to be young is as I grew older my knees and back had a harder time carrying all of the weight.

When I started working in the Brooks Range at the ripe old age of 38, I was carrying so much survival gear my pack was approaching the 60-pound weight. The Brooks Range is pretty isolated and the nearest settlements are hundreds of miles away. So, I wanted to be prepared if something happened to the helicopter and I was forced to rough-it for awhile. However, carrying so much survival gear had its downside. Being in my late 30s, my knees couldn't handle the weight and I ended up seeing an orthopedic surgeon after the field season. Because of the need for all of the survival gear in the Brooks Range, the BOM decided to make up survival packs of food, sleeping bags and tents and drop them at locations where a crew could easily walk to if they were stuck in the field. This eliminated a lot of weight from the packs. The survival packs were used one time during that field project when bad weather prevented the helicopter from returning to camp. I was not working with the crew at the time, but I was glad I was involved with the decision to have the field crews carry those survival packs.

In Alaska, a gun is a necessity. The guns issued to my field crews were either shotguns with slugs or .44 Magnum revolvers. I tried to carry a shotgun when in grizzly bear country. The shotguns would be carried with a sling and the revolvers would be in shoulder or hip holsters. Another apparent bear deterrent was pepper spray or capsicum spray. When sprayed in a bear's face, it causes tears, pain and temporary blindness. It also works well on humans because one year a USGS geologist with her Ph.D. decided to test out the effects during pepper spray training. As her fellow geologist was demonstrating using the spray, she stepped in front of it and got it full in the face. It knocked her to the ground screaming and writhing in pain. We all hoped it would do the same thing to a bear at least long enough for us to run away.

Clothing consisted of a couple of pairs of socks, sturdy pants, (Carharts or Filson were preferred), an undershirt (short or long-sleeved depending on the weather), long-sleeved shirt and usually a baseball cap. Since after the first couple of weeks, I could fit right in at a homeless camp, with my clothes being continuously sweat soaked and filthy, I didn't take very nice clothes into the field. An exception was a geologist I worked with who would wear his old dress shirts in the field. It was always kind of shocking to see a guy dressed in a long-sleeve, button-down white dress shirt in the field. I usually kept nicer clothes in camp, so I could change into them after a day of fieldwork. Clean clothes are a relative term when a person lives in a field camp that either has dust or mud, depending on the weather. One time I took a shower and put on my clean clothes. While I was writing a letter to my wife, I swatted a bug on my pants and a cloud of dust emanated from the slap site. Polypro clothing was a great invention. It is light-weight, warm and wicks moisture away from the body. Prior to this, I carried a lot of wool.

Carrying rain gear in Alaska is a must. I was lucky because prior to my first summer in Alaska, I went to a mining supply store in Denver and bought rubber boots and rain gear worn by underground miners. However, many people I worked with were not so lucky. They tried the early "Gore-Tex" and cheap rubber and nylon rain suits. This raingear works great on a cruise ship or vacation, but once the early Gore-Tex got dirty it leaked. It would usually only take one day of busting brush in the rain to trash the cheap rubber and nylon rain suits. After my mining rain gear wore out, I bought Helly Hansen rubberized rain gear, the kind used by commercial fishermen. It was heavy, but it kept me warm and wet. A person who hasn't worked in this rain gear would think the gear should keep a person warm and dry. But I challenge anyone to try to work in this rain gear and not sweat as they walk up and down mountains in the rain. However, warm and wet was preferable to cold and wet.

Clothing depended on where in Alaska I was working and the amount of experience I had accumulated over the years. When I went to Alaska in 1977, my only knowledge of Alaska was it was really far north and very cold. Therefore, armed with this logic, I bought long underwear and packed a down jacket. I carried those around with me all summer and never wore them until I was back in Anchorage and it snowed. When I worked in Prince William Sound for 3 seasons, the weather was mostly cold and rainy. During July and August of 1981, it rained 67% of the time. I wore a lot of long-sleeved shirts during that project. When I worked in the interior, many of the summers were hot and dry, so when I was working on the mountain tops I could get by with a t-shirt. During a field project in Goodnews Bay, I had to have winter gear because the summers were so cold and wet. It was so miserable out there I had Nathan Rathbun buy each field member full-face stocking caps. I went back to Alaska in 2000 and had a small project in the Alaska-Juneau Mine. The Alaska-Juneau Mine was being closed by an old BOM employee who I knew. He took my crew a mile underground, so I could teach them about mining methods, mine safety and underground mapping. It was so cold in the mine, I ended up putting on every piece of clothing in my pack, including my rain gear just to try and keep warm.

One piece of "clothing" I used in interior Alaska, which included the Brooks Range was a head net or bug net. In some places, the bugs were so bad I have killed 100 mosquitoes in a single swat. The bugs were so bad in the Brooks Range if I was walking with the wind, I tried to walk a little slower than the wind speed, so the bugs couldn't get to any exposed skin. They would hover in front of my face but were never able to land. Take that you little b**tards.

Clothing used when riding in helicopters was nomex. Nomex is fire retardant clothing wildland firefighters and pilots wear. As safety became a greater concern with the BOM, the BOM decided in order to fly in a helicopter, a person must wear nomex clothing. So, the BOM purchased green fire pants, yellow shirts and flight jumpsuits for each field crew member. People could choose which one they wanted to wear. There were advantages to each. If I wore the pant and shirt combination, I could wear the pants as my field pants and the shirt would go over any shirt I was wearing. If I wore the jumpsuit, I wore my field clothes and put the jumpsuit over my field clothes prior to entering the helicopter and removed it after landing. The advantages of the pant and shirt combination were I did not have to put them on before entering the helicopter nor take them off after leaving the helicopter unless it was so hot I didn't want to wear the shirt while doing fieldwork. This was very convenient when wearing rain gear. The advantage of the one piece jumpsuit was it gave me extra clothes I could carry with me. This was advantageous when working in the Brooks Range because I didn't have to pack extra clothes in case the weather got bad or I got stranded. Since it rarely rained while working in the Brooks Range, the jumpsuit worked very well. The disadvantage was I had to remove my boots when I put on the jumpsuit to get into the helicopter and I had to remove my boots in order to take off the jumpsuit when I got out of the helicopter.

There was always a shake-down period at the beginning of a field season. I and most everyone else who has gone out into the field has had the same experience. The first day out, I am constantly looking for things I forgot, especially little things. I would always remember my pack, rock hammer and rain gear, but might forget extra sample bags or magic markers. Once I got back to field camp, I would load up my vest and pack with all of the things I had forgotten to pack. After a couple of days of fieldwork, I usually had everything I needed to get the job done for the rest of the season.

Shakedown is different than lost items. A person could probably find out where geologists have been by doing a magnetometer survey over an area, looking for lost rock hammers. I would tie fluorescent flagging around my rock hammer, so if I set it down, I could easily find it. But, flagging still didn't prevent me from losing rock hammers every now and then. Because of the high cost of helicopter time, it didn't pay to go back to a place looking for a $20 rock hammer. Before I wizened up to the fact I was going to lose things in the field, I carried nice things. But, after losing a Buck knife and a watch, I only carry cheap items in the field. It's amazing once I started carrying cheap items, I never lost any of those things, but I still lose hammers. When I am scrambling around steep slopes or bending over and hitting rocks with a hammer, things are going to fall out of my field vest. So, I always made sure items like hand lenses, altimeters or extra compasses were tied onto my vest because things like pens, pencils, rulers and markers were always going to fall out and get lost. Luckily I always packed a lot of extras of those items.

I would be remiss if I didn't mention some of the high-tech equipment some of the modern field geologists use on a routine basis. Of note, all of this equipment has been developed in the last 20 years, after I left Alaska. Most geologists carry a GPS as either a hand-held unit or an app on their mobile phone. A GPS uses satellites and will give a position as either latitude-longitude or UTM coordinates. It has become a standard piece of equipment for anyone going into the field. GPS units range from basic models that just read a rough location to elaborate models used to map at a centimeter scale.

Some geologists map on either a laptop computer or a tablet computer. The computers are equipped with digital maps or aerial photos and can be outfitted with a GPS, so they can pinpoint their location on the maps or aerial photos. At the end of a field project, the data mapped on the computer can be downloaded into a digital mapping program like ARCINFO. Using computers for fieldwork really speeds up the generation of a finished product.

I have worked with geologists who not only use their cell phones to pinpoint their location using a GPS app, they also use the phones as a compass and altimeter. There's an app for those functions. However, GPS and cell phones do not work in underground mines.

A piece of equipment I have found useful for determining distances is a handheld rangefinder. Rangefinders have increasingly become more accurate and affordable over the years. A geologist can use a rangefinder instead of dragging around a 100-foot long measuring tape or trying to pace off a distance.

A piece of equipment used mainly for abandoned mine studies is a hand-held chemical analyzer. These machines give real-time results while in the field for a variety of elements. The only problem is they are very expensive and each has their specific limitations, which a person should be aware of before purchasing and using a unit.

A final note. All of the high tech equipment can be used by field geologists as long as the geologist is aware of its limitations and has the proper support behind them. Nothing makes up for a geologist knowing how to read a map or a compass because electrical powered equipment has a tendency to fail when a person most needs it. It's happened to me more than once. A geologist should also know their pace, which is the distance of each of their steps, in case their rangefinder quits or their tape breaks. Much of the high tech equipment and supporting software is expensive and requires training in order to use it. The geologists I know who use this equipment usually work for a business or government agency with the finances to purchase and maintain the equipment and train its personnel on its use.

Getting ready to take bottom samples in the Bering Sea at 10 PM

Chapter 10

Work Hours

My mama always said, "If a person wants a 9 to 5 job, don't become a field geologist." She didn't actually say this because I don't think she ever really knew what I did for a living. One of the things I learned from my first geologist job was there are no set hours and there was a limited number of days for fieldwork. In Alaska, there is a 3 or 4-month window, depending on where a person works. If I worked in the interior, fieldwork usually started in June because the ice in the rivers sometimes doesn't go out until late May. If I worked in southern Alaska, fieldwork usually started in late June or early July, because southern Alaska is more mountainous than the interior; therefore, the snow usually doesn't melt until late June. Usually, by September "termination dust", which means snow, halts all fieldwork.

Another determining factor on the length of a field project was the cost of doing fieldwork. Expenses associated with equipping a camp, mobilizing and demobilizing those camps, hiring crews, cooks and transportation are huge. Therefore, when I went into the field, I wanted assurance all of the areas I needed to evaluate were accessible. With so little time to get the work done and the large costs associated with the work, long hours and little downtime is the norm. Other factors contributing to the constant work ethos are the work is being done in isolated places, with few distractions, so a person might as well work. There is also the tough guy/gal persona. Finally, working long and hard is what geoscience peers expect of a field geologist.

Before working in Alaska, I think I was pretty typical. Being a twenty-something male right out of college, I was used to staying up late and getting up right before it was time to go to work or if I didn't have to go to work, I would sleep until 10 or 11 in the morning. These were the carefree days of being a grunt with no responsibilities besides humping a pack and beating on rocks. I followed this habit when I went to Alaska until Tom, who took over for Chris as manager of the WGM field project I was on, pulled me aside and said as the project geologist I was expected to be the last one to bed at night and the first one up in the morning. He also posted a sign in the office tent which stated, "I hired you, I can fire you, so get to work." This was in contrast with Chris's management style, which was more laissez-faire. After our little talk, I got in the habit of getting up between 5:30 and 6 a.m. every morning of the week. This habit has stayed with me into retirement. So, when the day started in Alaska, which doesn't mean daylight because it's usually kind of light all of the time, I would get up, get my field clothes on and go get breakfast.

I can cite an example of a typical field project work schedule because I kept records of the hours and days worked when I was on a project in Goodnews Bay in 1985. I spent 6 weeks helping the project leader, who was out of the BOM Fairbanks office, evaluate the beach and offshore platinum potential of the Goodnews Bay area in southwest Alaska. Goodnews Bay, specifically the area around Red Mountain, has been mined for platinum since the 1920s and still has the potential for producing more platinum. During the 6 week study, we sampled the beaches and the project leader took samples from bottom sediments in the Bering Sea off of Red Mountain. Platinum and gold were found in most of the samples. Since the BOM didn't have enough money to definitively sample the area, all the BOM could conclude was platinum existed on the beaches and offshore. Us government-types were always being disparaged about how cushy our jobs were and my records might help dispel any such notions, at least as it related to those government employees who worked for the now-defunct BOM in Alaska in the 1900s. In 42 days, I got 1 day off. A typical day would start a little after 6 a.m. with me getting up, getting dressed and making breakfast, usually an egg scramble. By 8 a.m. we would be out in the field. We would usually work until 8 or 9 p.m. The project leader would usually work longer because he had to write up daily work reports and figure out what we were going to do the next day. The day off came when the project leader left the project and went back to Fairbanks, leaving his co-worker behind to manage the project. The co-worker gave us a day off and he put us on a more normal 10-hour work schedule for the remaining 2 weeks of the project.

For this platinum study, the project leader wanted to take samples from the bottom of the Bering Sea. He did it by employing motorized rafts, diving down to the bottom of the sea with scuba gear and shoveling material into a bucket. We would haul the full bucket up into the raft and send down an empty bucket. During the project the leader decided that the sea was too rough during the day, so we would go out at night when he thought the sea was calmer. Therefore, we would start about 11 p.m., go down to the beach and launch two rafts through the surf. The leader would dive to the bottom as another worker and I would bob up and down in the 3 foot high swells of the Bering Sea. As we were riding the swells at 1 a.m., we would complain to each other that we were doing a really stupid thing. It would have been even worse if the show "Deadliest Catch" was on television at the time and we had seen how bad the Bering Sea could get. But, when the project leader came up off the bottom and asked us whether or not he should continue sampling, we said, "Yeah, why not, we're out here anyway at 1 a.m., might as well get the work done."

Working long hours is fairly normal for most field projects I've been involved with during my career, especially if I was responsible for my own food and not reliant on a cook. When a camp relies on a cook, eating times need to be pretty rigid or the cook is not going to be happy and no one wants an unhappy cook. My records indicate during the Kantishna study, breakfast was at 7:30 a.m. and dinner was at 7:30 p.m.

Thinking back, I didn't really get a day off the whole summer of 1974 when I worked for the USGS and the summers I worked on my thesis. In 1976, I worked 10 days on and 2 days off in southern Colorado. For WGM, I was the first one up in the morning and last one to bed at night, per my boss's remonstrance. For BLM, there were long field days because fieldwork was either by boat or truck and it took a long time to get to and from places. Then there were the years with the BOM where there was so much to do and so little time to do it.

Other field geologists can always top my stories. I

know geologists who work or have worked around the world and tell their experiences. It might just be a North American thing because the biggest complaint I hear from geologists who have worked in foreign countries is the lack of a work ethic which is prevalent with foreign geologists. It seems a lot of foreign geologists don't want to work as long and as hard as the North American geologists. It might be a cultural thing. It also might be a lot of North American geologists are in a foreign country for a short amount of time. They want to get the work done and get back home in the shortest time possible. However, there is the possibility North American geologists just have a screwed up work ethic compared to the rest of the world.

A person would think with all of the work geologists do and all of the hours they put in, they would be well compensated. The monthly salary for a field geologist is OK, but it is a monthly salary. Overtime pay is very rare, so if the dollars per hour are calculated, geologists are grossly underpaid. I guess geologists get paid for the satisfaction of a job well done or a job done well.

One compensation about working and living in Alaska was I knew, no matter how hard and long I worked in the summer, the dark and snow were going to descend and 8 months of office work was awaiting my return. Of course, working in the office also required extra hours. Coming home, I had to reconnect with my loved-ones, put away all of the field gear, deal with any samples that had been collected, compile all of the field notes and daily reports, deal with chemical analyses as they came in, draft maps, write reports, report on any findings, develop future plans for field projects and try to sell those plans to management. Then in the Spring, after everything was taken care of from the previous field season, it was time to get ready for hopefully another field season. The good thing about office work was I could usually set my own schedule. But I always knew ready or not the sun would ascend and the snow would melt and another field season would begin. Needless to say, I'd better be ready.

Staying in Alaska in the winter is not always an option for geologists who worked for private industry. Many of the industry geologists I knew would be assigned field projects in warmer climates during the winter; therefore, for them, it was the never-ending summer.

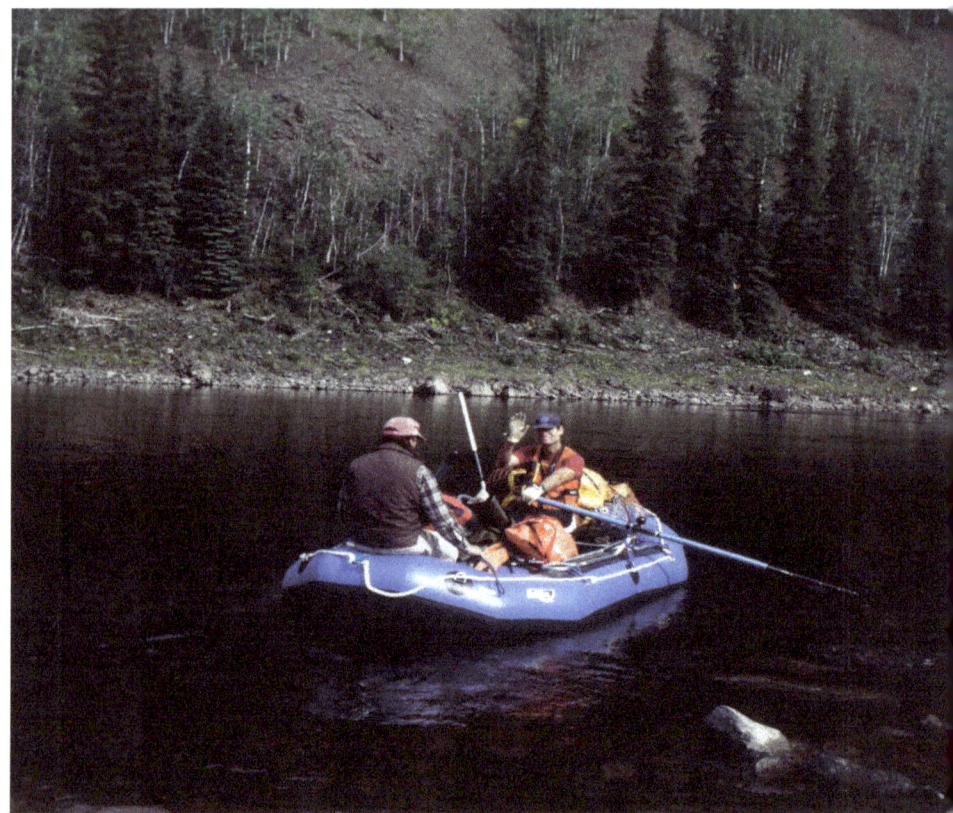

The author and Nathan Rathbun floating the Fortymile River in a raft

Chapter 11

Transportation

Ah, I love the smell of burning jet fuel in the morning. If a person has ever been around helicopters, they can relate because associated with the smell is the thrill of flying off and having some kind of adventure.

Transportation to the field in Alaska is eclectic. Some projects could be accessed via road. These projects were very rare and located largely in the southeast half of Alaska. Field locations accessible by road were in the Kenai Peninsula, Peters Creek near Talkeetna, White Mountains near Fairbanks, Kantishna in Denali National Park, and Porcupine Creek near Haines. We drove to the field locations with 4-wheel drive vehicles, from pickup trucks to SUVs. When we got to a field camp we would use other modes of transportation to access the field sites. Airplanes were also used to transport personnel and gear to remote field camps.

The preferred mode of transportation in field camps was the helicopter. I used various kinds of helicopters. One was the Soloy Hiller used on the Kenai Peninsula and in the Talkeetna Mountains. It was a turbine-powered helicopter capable of carrying 2 people in a bubble cockpit. Field gear was strapped to baskets on the skids. The advantage of the helicopter was it had a lot of power. The disadvantages of being slow and only able to carry 2 people, outweighed the advantage. Joe Kurtak and I had an interesting ride in a Soloy Hiller. We hired the helicopter, which was based in Palmer, to drop us off and pick us up a couple of days later in the Talkeetna Mountains. Everything went as planned. When the helicopter picked us up, the pilot got out and put a stick in the gas tank. I guess the helicopter's gas gauge was either non-existent or not working. As we flew back to Anchorage, the pilot kept on looking at the stick and muttering to himself. It wasn't reassuring to Joe or me because we assumed the pilot was trying to decide if he had enough fuel to make it back home. I guess he decided correctly because we made it back safely.

My favorite helicopters were the Hughes 500s. There were the C and D models when I worked in Alaska. The C model had less power than the D. Since the D was the newer model and therefore more expensive to rent, we usually got the C models. Hughes 500s have an egg-shaped cockpit. They were used during the Vietnam War. The shape of the cockpit supposedly was designed to withstand a crash. It was the helicopter of choice for field geologists because it was relatively fast, was little compared to all of the other helicopters, sat high off the ground and could carry 3 passengers. In some instances, the rear seats were removed and 4 geologists (seasonal employees) would be stacked in the back facing front to back.

Occasionally Bell Jet Ranger 206 helicopters were hired for fieldwork. They are relatively larger more comfortable and sit lower to the ground than a Hughes 500. Because they sat so low to the ground and had large rotor blades, they seemed to be more suited to an urban environment than the bush. But, if work was in an area with few trees, they were OK. A stretch version of the Jet Ranger, called a Bell LongRanger was used in the Brooks Range. It could carry 4 people comfortably. It was ideal for tundra covered areas because it was too big to fit into tight landing areas. Even though it was a more expensive helicopter to use, it was cheaper and safer in the long run because it could carry the whole field crew of 4 geologists to and from the field in one trip. A wonderful helicopter used for the Kantishna study was the Messerschmitt-Bölkow-Blohm Bo 105. This is a twin-engineer helicopter commonly used by hospitals for medivacs. It has a large storage compartment in the back. It is a great helicopter but was usually too expensive for most of my projects.

Boats were used in some projects. Alaska ferries were utilized to go from point to point in Prince William Sound. Rafts were used to float rivers when I worked for BLM and during the BOM Fortymile River abandoned mine study. They were also used in 1985 at Goodnews Bay to sample the bottom of the Bering Sea. On the Fortymile River, the rafts had rowing frames and were oar-powered. Motorized rafts were used in the Bering Sea.

During the 1985 platinum study at Goodnews Bay, the project leader contracted a supposedly sturdy boat. He used the boat along with motorized rafts to dive and sample the bottom of the Bering Sea. The sturdy boat consisted of two skiffs on which a local had nailed plywood in order to make a home-made catamaran. The local then built a wheelhouse on the plywood and mounted an outboard motor. It survived the Bering Sea while we worked off it, but I found out later it didn't survive the winter storms. During this part of the project, while the project leader was diving off this boat, he sent me and another field hand to Platinum to get some supplies. We took a raft and motored into Goodnews Bay and ran aground in the middle of the bay. How were we to know the bay only had a skim of water over it at low tide? The two of us just sat in the raft twiddling our thumbs for a couple of hours waiting for high tide. I'm sure our predicament was the talk of the town and maybe all of Alaska by the end of the summer.

The BOM also had a flat-bottomed riverboat. It was outfitted with a 45 horsepower Suzuki outboard motor and Nathan Rathbun had constructed a plywood cabin that provided a little protection from the elements. It was used to access points in Prince William Sound. Since the boat was flat-bottomed, no one wanted to take it out in the ocean unless the surface of the ocean was relatively flat. There weren't many calm days when I was working out of Whittier.

The riverboat was also used on rivers when fitted with a jet unit. Nathan Rathbun and I used the boat for the Yentna Mining District study. We were investigating the Susitna River drainage from Talkeetna downstream to the mouth of the Yentna River, then up the Yentna River. Nathan had a jet unit installed on the motor prior to this trip because we knew we would run into some shallow water. A jet unit essentially sucks in water and expels it out the back, providing propulsion. Nathan, I and a co-worker hauled the boat to Talkeetna. We launched the boat into the Susitna River and did a shakedown trip up-river. We did this because we knew we could float back down to Talkeetna if there were any problems. Every-thing worked great except we found out one of our gas tanks leaked. We decided to run the motor on the leaky tank first and repair it on the river when it was empty. We bid goodbye to our co-worker who would take the vehicle and trailer back to Anchorage and headed down the river. We took placer samples, camped out on river bars along the Susitna and started up the Yentna River on the 4th day. A half day up the river, we stopped at the Yentna River Lodge to say hello. After visiting awhile, we cast off and the motor wouldn't start. Luckily, we had packed a spare "kicker", which is a little motor that would only move the boat in calm water. We got the kicker working and were able to get the boat back to the lodge. We tilted the main motor out of the water and found the new jet unit hanging off the motor. The business that installed the jet unit had only secured the unit with 4 bolts instead of 6. When asked why only 4 bolts were used they said the unit only came with 4 bolts. They thought only 4 bolts were needed to secure the unit, even though there were 6 bolt holes. They thought wrong. To compound their mistake, when we requested replacement bolts via the lodge's radio, they sent the wrong ones. Fortunately, we had brought the original lower unit with the propeller that the installer had taken off the motor. We took the jet unit off and installed the lower unit with a prop, while standing in the river. Since we didn't want to risk damaging the prop and be up a river with only a kicker, we motored back to Talkeetna, avoiding any shallow water on the way. This little piece of fieldwork illustrates in Alaska a person not only has to be flexible but also has to plan for things that might go wrong.

All-terrain vehicles (ATVs) were great for some kinds of fieldwork. We used motorcycles, 3-wheelers and 4-wheelers. When I worked in Kantishna in 1983, the only multi-wheel ATVs available were 3-wheelers. These ATVs were the predecessors to 4-wheelers and were not very safe. They were like motorized tricycles on steroids. Because there was only one wheel in front, the front end was light. In order to prevent the 3-wheeler from flipping backward while ascending a hill, a person had to lean over the handlebars. There were a number of injuries associated with that phenomenon. And, because the ATV only had one wheel in front, if the rider turned too sharply, it tended to flip forward sending the rider flying. Luckily the BOM didn't own these machines because the ATV industry subsequently replaced them with much more stable 4-wheelers.

The BOM eventually bought four Suzuki 4-wheeled ATVs, which were used for specific projects. In the right situations, these gave my field crews a lot of flexibility because we weren't dependent on other people to provide transportation to the field sites. I used them in Goodnews Bay, where I didn't have the exclusive use of a helicopter but could access various parts of the district via limited roads and the beach. They were also used during the Valdez Creek and Fortymile abandoned mine studies where there were dirt trail systems.

One thing I learned during the Valdez Creek study was the need for a minimum of two 4-wheel ATVs when driving the dirt roads or trails in Alaska. Because of the presence of permafrost, the portion of a trail that crosses permafrost becomes a bottomless bog as the earth thaws out. The presence of a bog causes trail users to seek more solid ground by driving around the bog onto the tundra. However, if the new trail disturbs permafrost, then it too becomes boggy. The search for solid ground causes some parts of a trail to look like a braided stream with alternate trails going every which way. When we reached a braided trail area, the lead rider would roll the dice and hope to choose the most solid route. However, if the leader chose wrong, then they would literally get "bogged" down. Once bogged down it was almost impossible for an individual to extract an ATV from one of these mud holes because ATVs are very heavy. Therefore, every one of our ATVs had a tow rope. The stuck rider would throw their rope to their partner, who would drag them out of the bog. By the end of my work in Tyone Creek, I was well versed in ATV extractions.

Various aircraft used during the Colville Mining District study

Chapter 12

More on Aircraft

In May 1986, my crew and I were flying on a small commercial plane to start the Goodnews Bay Mining District study. Upon landing in Eek to let a passenger off, I saw the pilot reach for a broomstick he kept next to his seat, exit the plane and start banging on the wings with the stick. As he hit the wings with the stick, ice would fall off in chunks. This was his unsophisticated way of deicing his plane. The weather flying from Bethel to Platinum was cloudy and near freezing which caused icing conditions. Icing is dangerous because it adds weight to the plane and changes its handling. When the pilot got back in the plane he told us not to tell anyone. I was just happy to safely arrive in Platinum and figured what happened in the bush stayed in the bush.

Flying is to an Alaskan as driving is to a Californian. To get most places in Alaska, one must take some kind of aircraft. Airplanes have their advantages and disadvantages. They are faster, cheaper and carry more stuff than helicopters. So, they are very useful to travel from place to place. The disadvantage is they aren't as maneuverable as a helicopter. A helicopter can land in small areas, whereas an airplane needs hundreds of feet of clearance. A helicopter can hover and turn on a dime. An airplane needs room to make a turn. This was very important if the weather gets bad or the aircraft gets into a tight spot and the aircraft has to retrace its route. One important aspect of a helicopter is I also never got sick in one.

My worst experience in a small plane was when I first came to work for BLM. The guy who was in charge of the BLM fire program took us newly hired resource guys up in a plane, an Helio Courier owned by a local Tok family. The family owned an air taxi business and all of the pilots at the time were brothers. Much of their business was flying for BLM. When I moved to Tok, I flew commercially into Fairbanks and caught a ride to Tok with one of the brothers, who was in Fairbanks on BLM business. The family had a number of aircraft one being the Helio Courier, which is a STOL aircraft. STOL stands for short take-off and landing. These planes have large wings and can cruise at 28 mph. They are great in Alaska because they can land and take off in restricted areas. Resource people like them because they are cheaper than a helicopter and can be used to spot things like wildlife, fires and other activities. They almost have the same maneuverability as a helicopter. The problem with them is because they have such a large wingspan and cruise so slow, if there is any wind, they are like riding a bucking bronco. Well, the fire guy, who I think wanted to have some fun with us funny new guys, asked us if we wanted to see the resource area. We, of course, thought, "Wow what a great idea!" So, we went out to the local airstrip at Tanacross and climbed into this innocuous-looking aircraft. The flight started out OK, with a few ups and downs, but then I think the fire guy decided to have some fun with us. He came to an interesting place, at least to him. He had the pilot go round and round and up and down until we turned green. He finally relented and had the pilot land at the airstrip in Chicken, so some of us could call Ralph. I don't know if he was being nice to us or to the pilot and the interior of his plane. After the flight, the pilot confessed he really didn't like being a passenger in his plane because the only way he could not get sick in his plane was to fly it. Other than that one plane flight, I never felt ill in an airplane, probably because I just flew from place to place and never had to fly in another STOL aircraft.

I always tried to be vigilant whenever I was riding in an airplane. Joe Kurtak, Mark Meyer and I took a scheduled flight from Coldfoot to Fairbanks after finishing some work in the Brooks Range. The flight had one stop to make at the village of Anaktuvuk Pass before going to Fairbanks. I got to ride in the co-pilot seat and amused myself by watching the pilot go through all of the procedures prior to flight, the instruments during the flight and the scenery. The flight was very turbulent because we were flying over the Brooks Range, so I was concentrating on anything other than my churning stomach. I noticed the pilot didn't seem concerned with the plane pin-wheeling and bucking up and down. He just sat reading his book. Must have been a very good book because he was flying at an altitude where the plane may or may not have cleared the approaching mountain crest. I nudged his arm and pointed out the fast approaching ridge. He put his book aside, did some fiddling around with the gadgets in the plane and the plane rose a couple of hundred feet. The pilot then reopened his book and continued his reading. I don't know if I saved the plane from a crash or I was just being paranoid, but I was glad I didn't keep my mouth shut and didn't leave my future in the hand of fate.

Helicopters are more useful than airplanes, but they are big noisy machines with some moving parts that can hurt a person if he or she contacts those parts. While working around helicopters, I wasn't injured by big things, it was the little things. When I started with WGM in 1977, there was only one spare set of headphones in the helicopter. The person in the front seat would wear the headphones in order to talk to the pilot. The people in the back of the helicopter would just have to sit next to a whining transmission for up to a couple of hours at a time. I believe because of that summer in the field, I lost my high-end hearing. So, what was my loss of hearing worth? The answer is a couple of thousands of dollars out of my pocket for hearing aids, once not hearing gets more annoying.

When I started working for the BOM in 1980, the situation was no different. Helicopters did not carry headphones for each passenger. So, the BOM issued earplugs. I would take off my hat as I approached the helicopter, get in, buckle up and put in my earplugs, if I could find them and if they weren't too dirty or gross.

Big changes occurred when the Office of Aircraft Services (OAS), a branch of the Department of Interior (DOI) got involved with aircraft safety and contracting of aircraft for all of the DOI agencies. In the 1970s and early 1980s, they were mainly focused on safety and contracting aircraft used to fight wildfires. From the 1980s on, their influence over all of the DOI agencies' aircraft programs became more pronounced. DOI agencies had to use OAS to contract aircraft (helicopters and airplanes). OAS made sure an aircraft and pilot met certain standards. This did and probably does add costs and delays, but when an OAS contracted aircraft arrived on-site, I was reasonably assured it was an air-worthy craft and had a competent pilot. OAS also mandated safety training and got the agencies to adopt the use of personal protective equipment. They had yearly classes and also provided dunker training.

Dunker training simulates how to escape from an aircraft if it crashes into a water body. Since I flew over ocean, lakes and rivers, I felt it was necessary to understand what to do if an aircraft just happened to uncontrollably land on or in such a water body. Dunker training was conducted in a local high school pool. The trainers used a simulated cockpit, which was a cage of PVC pipes containing a seat with a seat belt and shoulder straps. The trainers strapped a participant into the seat wearing all of their clothes and flight helmet and tipped the cage into a pool. Even though the participant was immersed in heated water in a controlled situation, it was still disconcerting to find oneself under water, upside down and strapped to a seat. The participant, once they got over the shock of being underwater, had to undo the seat belt and shoulder straps and extract themselves from the cockpit. It was a powerful learning experience. Luckily, I never had to use the training, but I was glad to have that knowledge in my back pocket.

Personal protective equipment, which met some resistance by the agencies and personnel, included flight helmets, nomex (fire retardant) clothing, boot and gloves. This equipment added levels of confidence to the flying experience. Private industry geologists made fun of me when they saw my helmet and clothing. But, to me, it was like wearing orange field vests, not cool, but practical. Having my own helmet meant I would have some form of ear protection, even if I couldn't find my earplugs.

I have to apologize to any pilot reading this, but when

I was a passenger in a plane or helicopter, I regarded pilots as highly-skilled, glorified taxi drivers or chauffeurs. I wanted the pilot to pick me up, take me to where I wanted to go and drop me off safely. Most pilots did this competently. However, this was not the case with the first pilot I dealt with working for WGM. Most of the helicopter pilots working in Alaska at the time were Vietnam War veterans, who were used to flying in rough conditions. But, the WGM pilot who reported for work in 1977, got his flight hours flying a rich rancher around the San Joaquin Valley in California and as a pilot who flew tourists around San Francisco. Because he wasn't too experienced with landing where there were trees, bushes or mountain tops, he had a real tough time landing, especially in tight places. It seemed in order to get picked up, I would either have to be in an area with no vegetation taller than my knees or I would have to cut a football field-sized landing area, usually with my pocket knife. Since WGM's contract was only for a limited amount of time and therefore they wanted to get as much work done as possible, having their geologists spending time looking for or constructing suitable landing areas instead of taking samples was not acceptable. Therefore, after about a month, WGM transferred the pilot to a project where the tallest bushes were knee height.

Another pilot, who was working for another agency, liked to fly as if he was evading enemy fire. He would fly 120 miles per hour at tree height. Many of my "macho" compatriots thought the pilot was a cool guy. Since the pilot was working for those guys all I could do was make my opinion known to them, which probably lowered my macho-factor in their eyes. All I could think about when flying with the pilot was, "What will happen if the engine quits or we hit a bird?" He negated the best thing about a helicopter versus an airplane which is if there is a problem, like the engine quitting, the pilot can usually safely land the helicopter by using autorotation. According to Wikipedia, "Autorotation is a state of flight in which the main rotor system of a helicopter or similar aircraft turns by the action of air moving up through the rotor, as with an autogyro rather than engine power driving the rotor." This is a maneuver taught to all helicopter pilots and can be done safely if the helicopter is high enough off the ground so the pilot can find a good landing area.

I was involved with one autorotation in my field career. I was riding in the back of a Hughes 500 helicopter when the helicopter started shaking. The pilot cut power to the main rotor and autorotated to the ground. He checked out the helicopter and found the shaking was caused by loss of tape on the leading edge of one of the rotor blades. Tape is put on the leading edge of the rotor blades because the blades turn so fast rain can erode the blades if they weren't protected. The blades were also so finely tuned the loss of the tape caused the blades to oscillate and shake the helicopter. Once the pilot determined there was nothing mechanically wrong with his craft, he was able to fly us back to camp and get new tape for the blade.

Of all of the pilots I flew with, my favorite pilot was Ralph Yetka. He was a former Coast Guard helicopter pilot; therefore, he was used to flying in some of the toughest conditions. At the time he flew for us he was working for Temsco Helicopters out of Ketchikan, Alaska. He flew the Soloy Hiller and later Hughes 500s. He was a big man with a great personality. His only quirk was he preferred power-on landings or toe-ins to landing on flat ground. A power-on (my term) or toe-in landing is one used in steep terrain. The pilot flies the helicopter up to a relatively flat spot on a slope and rests the toe of helicopter's skid on the flat spot, while the rest of the skid is in mid-air. The passenger then exits or enters the helicopter by climbing on the skid. The passenger has to be aware of how their weight is affecting the stability of the helicopter, so climbing on and off a skid is done carefully. OAS did not like the maneuver, but for us geologists, it was either live with a little danger or be faced with climbing up or down hundreds or thousands of feet to get a sample, which in itself was dangerous. I would joke if Ralph had a choice between landing on a football field or on the goal post, he would choose the goal post. He felt as long as there was power to the machine, he maintained total control. He saved my crew a lot of weary sore legs and feet by doing toe-ins. I only worked one summer with Ralph but would get together with him whenever he was in Anchorage. Unfortunately, in 1987, he was flying a USGS geologist back to Ketchikan after a day in the field when a float plane took off from Ketchikan and crashed into Ralph's helicopter. The floats from the plane took out the helicopter's main rotor and it crashed into the sea killing everyone on board. The plane was able to fly back to Ketchikan.

I had a notable incident concerning a helicopter contract. OAS hired a helicopter (equipped with floats), a pilot and mechanic from the lower 48 to work for me in Prince William Sound. I wasn't too happy because I was used to using Temsco, who usually had pilots familiar with flying in Alaska. The contract did not start out too well when the helicopter showed up late. I wasn't a happy camper because the field crew was ready to go, it was costing about $1,000 per day to sit and do nothing in Cordova and I was already dealing with a short field season (just 2 months). When the helicopter eventually showed up, the pilot, explained they were delayed by weather. The helicopter did not look in the best condition. The mechanic had to blow up the floats every morning because they leaked. Since he and the pilot didn't bring an air pump, the mechanic sometimes blew up the floats with his mouth. Helicopters flying over large bodies of water, like Prince William Sound, need to either have enough altitude to be able to autorotate to dry land if they lose power or have floats attached to the skids. There are two kinds of floats: ones that are inflated all of the time, or the compact ones that inflate when they strike the water. The compact floats create less drag but are more expensive. It wasn't until later in the summer, when the pilot and mechanic felt more comfortable with us that we learned the real reason they were a couple of days late. The company who owned the helicopter couldn't get us the helicopter on time because they first had to deal with mechanical issues, like finding used floats. They also had some financial issues.

Financial problems caught up with the pilot and mechanic when I was moving camp from Cordova to Valdez. The day of the move the pilot and mechanic found their helicopter chained to the Cordova runway. The fuel company in Cordova was holding the helicopter hostage because the helicopter company had not paid their fuel bill. Luckily for me, I was moving the office and personnel via the Alaska ferry system from Cordova to Valdez, so I didn't need the helicopter for a couple of days. I had also received a second helicopter, which wasn't needed by another field crew, so the lack of one helicopter was not going to impact my field project. I felt sorry for the guys who were left in an uncomfortable situation, but there wasn't much I could do about it. The helicopter company finally paid their bill and pilot and mechanic were able to escape Cordova and join the crew in Valdez. Twenty years later, I met up with the pilot in Phoenix while he was working on another helicopter contract. Small world.

As the old saying goes, "Experience is achieved when you survive your mistakes". If it's not an old saying, it should be. During my second year of flying out of Cordova, I was working halfway between Valdez and Cordova. The mountaintops in Prince William Sound were socked-in, so I was working below cloud level. At the end of the day, the helicopter was running low on fuel and we decided to head back to Cordova. Prince William Sound is a series of fiords or inlets. The safest thing to do in bad weather was to follow the shoreline; however, the pilot didn't think there was enough fuel to follow the coastline. So, the pilot took a straight path across open water back to where he thought Cordova should be. This was long before the advent of GPS, so a lot of flying was done by dead reckoning or as they call it in Alaska IFR, which stands for "I Follow Roads or Rivers." In this incident, IFR was not going to be helpful. While we were flying, I was trying to read the map and looking for any sign of land through the clouds and mist. When we finally got to what we thought was the mainland, the pilot asked, "Right or left?" I thought I recognized the shoreline, so I said, "Left" and luckily I was correct and we made it back safe and sound. However, I gained experience that day and never got into that kind of situation again.

The closest I ever came to being involved in a helicopter crash was working out of Valdez on August 13, 1982, a Friday. Friday was sunny and hot, which was unusual for the Valdez area. Nathan Rathbun and I were examining prospects and were having a rough day. It was one of those days when I thought it would have been better to stay in bed, but who could stay in bed on a blue sky day. In the morning on our first prospect examination, Nathan and I had to tie a rope around a boulder and use it to get to a prospect. The rope wasn't for technical climbing but was used as a safety line in order to negotiate a slippery vegetated slope. After we mapped and sampled the prospect, we climbed up the rope and were congratulating ourselves for not ending up like the boulders we pushed down the slopes. It was then Nathan realized he had forgotten his rock hammer. So, I went back down the rope and got his hammer. Yeah, we survived this incident to map some more. We then went to the Gold King Mine. It was a high-grade gold mine. After mapping the surface, we went to the mine opening and found it plugged solid with ice, so we couldn't map and sample one of the best prospects in the area. In the afternoon, Nathan and I got dropped off on a ledge on the side of a mountain using the power-on landing maneuver. The helicopter then went to the valley floor to await our call. When we were finished mapping and sampling a prospect, we called the helicopter and Nathan and I made our way back to the ledge. When the helicopter came into the ledge, the rotor wash blew the maps from my clipboard down the slope. I did mention it was Friday the 13th. Nathan said he would pick up the maps if I would get in the helicopter. So, I snuck around the nose of the helicopter because the pilot was on the left side of the helicopter and I was approaching from the left in order to get in the right front seat. The pilot kept the power on because the ledge was narrow and he had to keep the main rotor away from the slope of the hillside by just resting the front of the skids on the ledge. I climbed into the helicopter and because it was hot, I just sat in the seat, with the door open and tried to figure where we were going next. Meanwhile, Nathan was approaching the helicopter with my maps in hand. The pilot, who had opened his door in order to cool off, yelled at Nathan asking him to close the pilot's door. Nathan couldn't hear him over the roar of the helicopter, so the pilot in frustration, finally put the cyclic, (the stick that steers the helicopter), between his legs and reached over to close the door. While the pilot was reaching and I was half hanging out of the helicopter, the cyclic slipped forward and the helicopter started moving toward the mountain slope about 10 feet in front of us. I sat there thinking, "Do I jump or ride it out?" I decided to ride it out and the pilot ended up grabbing the cyclic, pulling back on it and the helicopter leaped backward into the air. Once the pilot got his composure back and I got the door shut and was buckled into my seat, he flew the helicopter back on the ledge and picked up Nathan. Because the day had been full of unusual events and it was getting late, I made an executive decision and called it a day. Sometimes the better part of valor is to acknowledge defeat at the hands of fate or as another old saying goes, "Don't try to push the river."

Sometimes working through a third party, like OAS for helicopter contracting can have its downsides. Mike Balen and I were working in the White Mountains. We had a helicopter pilot who was kind of unstable. We suspected, but couldn't prove he was doing drugs. Mike saw him one day trying to balance the helicopter on top of the gold dredge near Chatanika. I found out later he also scared the other field members by flying below tree level in the drainages. I called OAS to see if they would replace him, but they told me because I had no proof of any illegal activities, they wouldn't send a replacement pilot. After my work was done, the pilot and his helicopter went to work for Joe Kurtak, who was the crew boss on the Valdez Creek project. Joe was finally able to replace the pilot when the pilot ran out of fuel while flying and set the helicopter on the tracks of the Alaska Railroad. The railroad was not happy when their trains were delayed until the helicopter could be refueled and flown off the tracks.

Finally, there is a job a helicopter pilot performs that requires more skill than just flying butts from point to point. The job is slinging equipment and supplies to remote locations. Moving supplies and equipment requires a cargo net and a line with a ring at one end. The procedure to sling equipment is: 1) a cargo net is laid on the ground; (2) equipment and supplies are piled onto the cargo net; (3) the net is drawn tight around the equipment; (4) a line is run through the cargo net; (5) the helicopter is flown over the net; (6) a person carries the ring, which is attached to the line under the helicopter; (7) the person puts the ring over the hook attached to the belly of the hovering helicopter; (8) the helicopter gently gains altitude until the load is off the ground; (9) once the pilot feels he can lift and fly the load safely, the load is flown to the remote site; (10) at the remote site, the load is gently lowered to the ground; and (11) the pilot hits the release button that frees the ring from the hook. The people at the remote site will unload the cargo net and usually pile the net and line into the back of the helicopter.

Over my career, I have helped sling camp supplies, tents, outhouses, fuel barrels, drill rigs, rock samples and other sundry items. I never liked slinging stuff for many reasons. There was a lot of downtime between sling loads depending on how far the stuff had to be moved. During the waiting periods, all I could think about was all of the geologic work that wasn't getting done. Pilots had to be careful about how the load moved through the air;

therefore, they usually flew pretty slowly. The last thing a pilot wanted was for the load to get wrapped around their tail boom or hit their tail rotor. A helicopter can only carry so much weight; therefore, I was always trying to maximize the load without overloading the sling. I hated it when the pilot released the sling because there was too much weight for the helicopter. I always seemed to get the job of hooking the load up, probably because I was tall. I would have to stand under the rotor wash with dust and debris flying around me and reach as high as I could to put the ring on the hook. I would always wear gloves because I would slap the helicopter hook prior to touching it with the ring to release any static electricity that might have built up on the hook. After hooking up the load, I would find a safe place to stand and signal the pilot it was a "Go." One thing a person has to remember if they are ever involved with slinging stuff is if the pilot feels like the load is either too heavy, nonairworthy or just doesn't feel right about the load, they will hit the hook release and send the stuff plummeting to the ground. Seeing equipment smashed on the rocks usually ruins the whole day.

Denise Herzog prospecting on ridges in the Valdez Creek Mining District

Chapter 13

Prospecting

"Slip-slidin' away, slip-slidin' away, the nearer your destination, the more you're slip-slidin' away," I sang to myself as I fought gravity to complete a traverse on a steep hillside. I wondered if Paul Simon, who wrote this song ever worked as a field geologist. Probably not, but the song was apropos for the situation I found myself in, and while singing it I had to contemplate the profession I chose that got me into a situation where with one misstep I might find myself sliding off a mountainside.

The definition of prospecting is the search of a region for something, but it is really just treasure hunting or looking for stuff. There aren't many grizzled prospectors crossing the country on foot, dog team or canoe anymore, but there are prospectors, who after a summer, look like cousins to a grizzly bear, crossing the country in helicopters, 4-wheelers, trucks and boats. Here's how this sometimes grizzled prospector did it and still does it.

First I gather all of the geology, mining literature and available maps of an area. The internet has made literature searches a lot easier. I then sit down and painstakingly read the reports and look at the maps to determine where to prospect. I look for any current or past mining claims and identify the known mineral deposits. Files of each property are made that include the location, summary of the property and copies of the discussion of the property in the literature. The geochemical data for each creek in the area is also reviewed and any geochemical anomalies worth a second look are identified. When I have identified the scope of the work that needs to be accomplished, a field program based on the work and available funding is designed. Some areas can be prospected in a couple of days and some take years. I have been involved with each situation.

Over my 18 years of fieldwork in Alaska, I did my share of prospecting for buried treasure. As I've previously said, I started out doing basic exploration work for WGM on their Doyon contract. The Doyon region is located in the central uplands and lowlands system. I wasn't involved in the first year (1976) of the contract. The work consisted of basic exploration. WGM sampled the streams in the area. Two samples were taken from each stream, one stream sediment sample and one pan sample. A stream sediment sample is taken by reaching down into a stream bed and trying to grab fine-grained material (dirt and clay). Handfuls of this material are put into a paper sleeve. It is hard on hands, fingers and fingernails. In theory, if there is a mineral deposit present in a drainage, the water that is leaching the mineral deposit will be rich in whatever element or elements are contained in the deposit. The fine-grained sediment in the drainage will adsorb the ions of those elements from the water. A stream sediment sample is taken on a river or creek and upstream from each tributary on the river or creek.

A pan sample is also usually taken from the same location. A pan sample consists of shoveling gravel into a gold pan. The contents are panned down to enough material that will fit into a small plastic bag. The heavy minerals from a pan sample are then analyzed. This method is used to see what minerals are eroding from a mineral deposit. It is a good method to find such things as gold, platinum, tungsten and tin. Both these sampling methods do not tell how much good stuff is in the creek or drainage, just that one creek or drainage might be a better prospecting location versus another.

My first prospecting job in Alaska was working as a project geologist for WGM in 1977. I was involved in the second year of the Doyon study. The 1977 work included evaluating the results from the 1976 sampling program, determining if the sample spacing was adequate enough to fully characterize the geochemistry of a drainage, mapping and sampling any known mineral deposits and prospecting for unknown deposits. Even though the 1976 crew had done a great job taking geochemical samples, I found in some cases, the sample spacing was not adequate to characterize a drainage or in other words, find the good stuff if it was present in a drainage. My job as project geologist was to determine if more stream and pan samples needed to be taken in drainages, then identify where the samples had to be taken to fill in the gaps. I would then give everyone maps indicating where to take samples, and then participate in days of frenetic sampling using a helicopter. I ended up spending a lot of time in 1977, with my hands and feet in the water and busting brush (i.e., walking).

Fill-in sampling involved stream hopping. Most helicopter pilots loved this work because it involved a lot of flying. Each geologist would tell the pilot where they needed to take the samples and the pilot would get them as close as possible to the sample spot. Sometimes the landing spot wasn't very close to the sample location and we'd have to get to the sample location as fast as we could through whatever terrain and vegetation was in our way. After the pilot dropped us off, they would go drop off the next person. The goal was the helicopter would never shut down, so we always wanted to get the sample collected as fast as possible and get back to the landing area before the helicopter returned. The only break we had was when the helicopter had to refuel. This fill-in sampling was kind of fun and stressful at the same time.

If the sampling was adequate and high or anomalous chemical values were present in the samples taken, the creeks would be walked in order to try and locate the source of the high values. This work was the same during my whole career in Alaska. The only difference between the work I did for WGM versus the BOM was the BOM had either the Alaska Division of Geological and Geophysical Surveys (ADGGS) or the USGS do the stream and pan sampling. The follow-up work was fairly similar in all cases. If the stream samples had high geochemical values, I would start at the headwaters of a drainage and walk down it, looking at the rocks and outcrops until I reached the mouth of the drainage. This was the method used if helicopter supported. If I was on foot or on a 4-wheeler, the drainages would usually be walked up. To walk up or down a drainage is a matter for debate among geologists usually over an adult beverage. It is generally agreed walking up is the best method because one's head is down, the rocks appear to be closer and while a person is resting during the climb, the person is looking at the rocks more closely. Also, if there wasn't any helicopter support, it was always easier to walk uphill in the morning when fresh and walk downhill to the vehicle or camp in the afternoon. However, after all this discussion, I contend walking down is easier, faster and if done carefully, just as productive in seeing the rocks as walking up. And, needless to say, it's a lot easier.

When prospecting, I look for anything that looks interesting. It was usually any rock with staining or quartz-veining. Many economic minerals like sphalerite (zinc sulfide), galena (lead sulfide) and chalcopyrite (copper iron sulfide), when weathered, stain rocks. Sphalerite and galena both have a white stain, copper has a green or greenish-blue stain and any mineral with iron can have a yellow or rust-colored stain. I would look for anything with quartz veins because gold and silver are commonly found associated with quartz veins. Prospecting can be frustrating when an iron-stained outcrop is found and is so weathered no fresh minerals are visible. When these kinds of outcrops are found, the only way to get a fresh sample of rock is to drill it, but there is usually no money available to get a drill unless something valuable is found first. It's a Catch 22. This happened during the Kantishna study where Bob Hoekzema and I walked a creek with yellow-stained acidic water. I was glad for rubber boots and gloves because I think the water would have dissolved leather and possibly skin. We walked up the creek and found an iron-stained very weathered outcrop. The creek and the weathered outcrop indicated there were probably sulfides somewhere below the surface leaching and forming sulfuric acid. We, therefore, suspected there should be something interesting at depth, but our contract with the National Park Service didn't allow us to do exploratory drilling. This mystery will remain a mystery until the government starts exploiting the mineral resources in the National Parks, which won't happen in my lifetime.

Prospecting in creek beds mostly involved "bustin' brush". I said walking down a creek is easy. And in some cases, like up in the Brooks Range where there are no trees or brush it is like a walk in the park, without a trail. However, in a lot of Alaska, walking down a creek means dealing with whatever vegetation Alaska has to offer. This can include alder, devils club, swamps, bogs, trees, and high water. Some of the worst vegetation is alder. Alder grows out from a bank or points down on a slope in patches. While walking along there were usually alder patches. To get across the alder patches, I would grab hold of one alder branch and swing from branch to branch, hoping the alder would hold and I wouldn't find myself in the creek.

There have been a couple of days walking creeks that tested my resolve. Some of those days were on Kayak Island in Prince William Sound. The USGS had taken stream sediment samples with high zinc values. The sample results were so high it seemed guaranteed there was a mineral deposit in the sampled drainages. I was so excited by the prospect of finding a new mineral deposit on Kayak Island I could hardly wait to walk the drainages. The first day my field partner and I helicoptered into the top of one of the drainages and Mark Meyer and his field partner were let off at the top of another drainage. The drainages were only one mile long, so I was looking forward to a quick productive traverse. Going downhill should have taken less than an hour. Instead, we encountered brush, waterfalls and devils club. Devils club is a plant with broad leaves on a stalk covered with little thorns. Those thorns penetrate the skin and fester. It is Alaska's equivalent to cactus. There are forests of devils club on Kayak Island. They towered over us and were up to 12 feet tall. It took 2 hours to investigate one creek. To make matters worse, nothing, but barren rock was found. The other field crew had the same experience in a parallel drainage. I, therefore, made an executive decision to not walk every foot of every remaining creek. The basis for the decision was there were a lot of creeks to prospect on the island, the creeks were difficult and time-consuming to walk and we had prospected the creeks containing the best values and found nothing. Spending a lot of time on Kayak Island seemed like a waste of precious field time. I decided the helicopter would drop a crew off at the top of a creek, we would walk down to the first cliff, then go back to the top. The helicopter would then pick up the crew and drop them at the bottom of the creek where we would walk up to the first waterfall. Because the drainages were so short and steep, if there was a major deposit hidden in them, there should be mineralized rocks in the creek bed. Since no one ever found any mineralized rocks in the creek beds, I was happy with my decision to not spend time being beaten up by nature. This sped up the prospecting of the drainages. It seems to have been a good decision because I talked with the USGS and they spent a lot more time than I did walking all of the drainages multiple times and did not find anything. The USGS finally concluded that metals in the stream sediments are derived from the weathering of a rock formation on the island. No ore deposit has ever been found on the island. I always pity the poor industry geologist who in subsequent years looks at the geochemical data and gets assigned to find the huge zinc deposit on Kayak Island.

Besides walking down creeks, slopes would be prospected, if I suspected something might be on the slope or there were stained rocks or quartz veining. This just entailed going for a walk, kind of. I once got a good lesson in walking on slopes in bad weather when working out of Cordova during my first year as a crew boss. I wanted to prospect across a slope along a fiord in Prince William Sound because it was suspected there might have been a hidden prospect on the slope. The helicopter pilot let me and my field assistant off on the top of a ridge and then flew to the bottom of the slope to wait for us. It was a day where clouds would skid across the tops of the ridges, but because we were going to go downhill, the cloudy conditions shouldn't be a problem. The slopes of Prince William Sound are covered by a plant commonly referred to as grease grass. It's not a grass, but it sure is slippery. Later on, we used to carry ice creepers to put on our boots when we had to traverse slopes with grease grass. Because the slope looked treacherous, I told my field partner to wait while I figured out the best route to take and started down and across the slope. After going a couple of hundred feet, my feet slipped out from under me. Slip-slidin' away. In my hands were a clipboard, rock hammer and shotgun that went flying when I slipped. I stopped my slide, collected all the stuff and started again to cross the slope. I slipped again and low and behold, there was a bear in my intended path. I decided because it was almost impossible to keep my footing on the slope and it was going to be tricky to get past the bruin, a hasty retreat was the best option. Therefore, I yelled at my partner to stay where he was because I was going to go back to the ridge, which I proceeded to do, slipping and sliding all the way. Once back on the ridge I called the helicopter. By the time the helicopter cranked up, the clouds came in and covered us. This happened for over an hour. Finally, when there was a sufficient break in the clouds to get "rescued" we went to a lower elevation and I decided tempting fate when the weather was bad was a pretty stupid thing to do. To this day, I don't know how we would have got off the ridge safely if the weather had gotten worse.

I found a letter to my wife about a traverse I took with Bob Hoekzema. It said we walked 1.5 miles in 8 hours while hanging onto everything we could find because the slope was so steep. I remember this particular day because I was even grabbing hold of devils club in order to not slide down the hill and spent the night picking thorns out of my hands and thighs.

Following up anomalies can pay off. An example of this was when I was working in Porcupine Creek near Haines. Porcupine Creek has been made famous by the Discovery Show, "Gold Rush: Alaska". Many years prior to the show, Bob Hoekzema, Nathan Rathbun, Gary Sherman and I were evaluating the placer deposits near Haines, Alaska for the BOM Haines-Klukwan study. Placer gold is found in the gravels along Porcupine Creek. We took samples from gravel deposits as we walked up the creek. We noted the gold was rounded near the mouth of the creek and as we went upstream, the gold became less rounded and eventually looked rough like it just came out of the vein. While sampling, we walked by some small quartz veins in the walls lining the creek. After we passed the veins, the gold values decreased markedly. Therefore, through our sampling, we were able to isolate these small quartz veins as one of the sources of gold in the creek. Of note, when gold is eroded from its source, it is flaky. As water moves it down a drainage, it is pounded by the gravel in the drainage and the edges of the flakes start to curl. This curling continues during the travels downstream until a roll or nugget of gold is formed.

The Red Dog lead-zinc deposit in northwestern Alaska was discovered by following up a geochemical anomaly. In the mid-1960s, Dr. Irv Tailleur of the USGS was fly-

ing over northwest Alaska when his pilot said he had noticed a red creek Irv might be interested in investigating.
Irv told the pilot he would check it out. So, the pilot
landed near the creek. Irv took a stream sediment sample and some samples of rocks in the creek bottom. Irv
got back in the plane flew off. Dr. Tailleur published
the results of sampling of the creek along with the other
creeks he investigated in the area. In the 1970s, the BOM
contracted WGM to evaluate the area surrounding the red
creek. WGM noted the high values of lead and zinc in the
stream sediment sample Dr. Tailleur took and went to the
creek to see what was causing them. WGM found a bed
of almost solid galena and sphalerite 4 feet thick in the
stream. After the results of the study were published by
WGM for the BOM, the area, which had been withdrawn
for a national park/monument was reopened to mineral
entry and selection by native corporations. The red creek,
which was named Red Dog after Irv Tailleur's pilot's dog
was developed into a world-class lead-zinc mine. A con-
troversy arose while I was working for the BOM as to
who was the discoverer of the deposit. Was it the pilot who pointed out the red creek, the USGS, who took a
stream sediment sample and published the results, WGM
who actually found the mineralized outcrops or the BOM
who contracted WGM? After much wrangling and pos-
turing, it seemed the mine operator, Cominco, couldn't
take the bickering and finally settled the matter in the
late 1980s by just saying that Red Dog was discovered
by the pilot.

Mike Balen taking a placer sample
using a mini-sluicebox

Chapter 14

Prospecting Techniques

I use and have used a number of techniques to prospect for hidden treasure. A technique unique to the BOM in prospecting an area was the use of a large placer sample. Bob Hoekzema pioneered or invented this method. Taking a pan sample of gravel in a creek will not give a good indication of how much gold may or may not be in a creek. If the sampler is lucky, a pan sample might indicate gold in the creek, but little else. There is also a good possibility the sample would be taken where there is no gold. Bob looked at the statistics surrounding placer sample sizes. The statistics show a curve, with the margin of error being great when a small sample is taken and the margin of error decreases as the sample size increases. The curve has an inflection point where the margin of error goes from large to small. The margin of error never does reach zero. As one would expect, the margin of error is huge with a sample the size that fits in a pan, and the margin of error doesn't start dramatically decreasing until 16 pans are taken. Sixteen pans are equal to about 0.1 cubic yards of material or around 300 pounds.

When I joined Bob's crew in 1980, our job was to evaluate the Kenai Peninsula and the northwestern portion of Prince William Sound for the Forest Service's RARE II study. This area was noted for its lode and placer gold deposits. Bob decided to employ the technique of taking 16 pan samples from one location. It worked well. Every drainage was sampled at intervals that depended on the size of the drainage. The technique was simplified when Bob added a portable sluice box. Portable sluice boxes are about 3 feet long and 10 inches wide. The box is lined with carpet, with expanded metal and riffles on top of the carpet. The sluice box is placed in the bottom of the creek, so water flows over the riffles. One person shovels gravel into a pan and hands it to the person manning the box. The sluice person dumps material into the head of the sluice while the other person loads another pan. Using a sluice box mimics how a miner recovers gold in a placer plant. I used this technique in most of the areas I studied because it was a relatively fast method that gave good results. After taking a 0.1 cubic yard sample, I was fairly confident a drainage did or did not contain gold.

The placer prospecting technique was used effectively in the study of the Valdez Creek Mining District. In the southern part of the district, miners told me they liked to mine creeks that drained a conglomerate bed on top of the ridges in the area. A conglomerate is gravel that has been hardened into a rock. The miners suspected the conglomerate was the source of the gold but was not economic to mine. So, I took a field crew with 4-wheelers into the area. We were able to drive 4-wheelers on top of the ridges and fill buckets full of decomposed conglomerate. We then processed the conglomerate through a portable sluice box. We found gold and platinum in the conglomerates and proved the miners were pretty smart. The concentration of gold in creeks that drain conglomerates is also very common in the Peters Creek area, west of Talkeetna.

Probably the stinkiest sampling occurred when I worked in Prince William Sound. Salmon would be spawning in many of the creeks we were sampling. Don't get excited, we only shoveled gravel from above the water line. Because of all of the salmon in the creek, we would have to move them out of the way in order to place the sluice box into the creek. Then we would have to try to keep the salmon from swimming up the sluice box. All the time we were working we were surrounded by spawned out salmon in the creeks and dead salmon half eaten by bears and birds. I was always looking over my shoulder to see if some bear decided to return for a mid-morning snack while we were taking these samples.

To compound the unpleasantness of sampling salmon spawning creeks, my lunch of choice was usually some kind of canned fish and crackers. Chowing down on fish while watching fish spawn and rot wasn't too appetizing, but I always soldiered on. Why did I take this food into bear country? Cans usually hold up to the abuse of cramming rock samples into a pack and they contained a good dose of protein and calcium. I also figured a bear would probably be more attracted to the 210 pounds of tasty meat standing before him rather than an 8 ounce can of fish.

The best placer sample I ever took was near the village of Goodnews Bay. While sampling a creek draining Granite Mountain, Wattamuse Creek, Bob Hoekzema and I found placer gold beneath every rock we turned over. Usually, a placer sample would only contain gold particles as big as fly s*it, but this sample contained mostly coarse gold. The sample from the creek contained 3/4 of an ounce per cubic yard. There is a tragic end to the story. Wattamuse Creek had been claimed by a miner and his wife. The miner and his wife lived in the village of Goodnews Bay and had applied to the government for a patent to the claims. At the time, if a miner successfully navigated all of the regulatory requirements, their claims could become private land. This miner officially filed all of the paperwork and patiently waited for the Federal government to act. During the 5 years of waiting, the miner died, leaving the claims to his wife. The wife didn't think she had to do her annual assessment work because she had filed for a patent. However, not filing the assessment paperwork is a fatal flaw that makes claims null and void. Not only did she lose her claims, but the land had been withdrawn from mineral location, so she couldn't stake a new claim over her mine site. She called me asking what she could do and all I could tell her was her only recourse was to try and tell her story to her Congressmen who might have more pull than I had in dealing with Federal bureaucracy. I don't know if she followed up on my suggestion.

I used other prospecting tools besides placer sampling. "Zinc zap" was used extensively when I worked for WGM. Zinc zap solution consists of mixing equal parts of Solution A and B together.

Solution A: 3% potassium ferricyanide in water Solution B: 3% oxalic acid, 0.5% N,N-diethylaniline in 1% HCl (i.e., 9 ml concentrated HCl, 30 g oxalic acid, 5 mL diethylaniline in 1 L water)

WGM would issue spray bottles containing this solution, which has a pretty blue color. If anyone saw a white coating on a rock and wondered if the coating was zinc oxide, they would spray the zinc zap on the rock.

If there was zinc, the coating would turn from white to red. The use of zinc zap was discontinued because of the toxic nature of the solution, but when it was in use, it was used liberally because it was such a good diagnostic test. There might still be a lot of blue outcrops in the Alaskan wilderness.

Another chemical all geologists carry is a weak solution of hydrochloric acid. Acid is put on any rock that might contain carbonates, like limestone, marble and dolomite. If there is carbonate in the rock, the carbonate will fizz when hit by the acid. Many mineral deposits, like the gold deposits in Nevada and the lead deposits in Missouri, are found associated with carbonate rocks, so I was always on the lookout for rocks that fizzed.

A chemical test used to field test for tin is called the "tinning test". To conduct this test, all that is needed is a source of zinc and dilute hydrochloric acid. Since every geologist carries hydrochloric acid, all that is needed is zinc. Nathan Rathbun had some zinc bowls made from a 3-inch diameter rod of solid zinc. The bowls were about 1 inch deep. These bowls were used to evaluate the White Mountains for their tin placer potential. A placer sample would be taken, the concentrates would be panned and some of the concentrates would be put into the zinc bowl. Acid would be added to the zinc bowl and if there was tin in the form of the mineral cassiterite, then the cassiterite particles, which are normally a dark color, would have a silver (tin) coating on them.

Another prospecting tool was a black light. This was used when the tungsten mineral scheelite was expected. I carried a black piece of visqueen with me in areas that might contain tungsten deposits. I would take either a rock or the placer concentrate, crawl under the visqueen and turn on the black light. This was especially uncomfortable when it was hot and there were a lot of bugs. If scheelite is present, it glows blue under a black light. Some geologists would wait until nighttime in order to evaluate deposits where scheelite was suspected, but since everyone is working in the land of the midnight sun, waiting for "dark" is sometimes a long wait.

There are other field tests that provide a good indication of what minerals are present. Field tests for gold and platinum are fairly easy. Gold is soft, malleable and heavy. I use a knife to tell the difference between gold and fools gold. Some minerals like pyrite and chalcopyrite look somewhat like gold, especially when there is no gold around for comparison. But a knife blade can be used to crush each of these minerals into a powder, whereas gold will just smear.

Platinum is silver in color, heavy and harder than gold. There are things that might be confused for platinum, such as steel and lead. However, it is easy to tell the difference between these metals because most platinum minerals are not magnetic or are weakly magnetic, whereas steel is highly magnetic. Therefore, a magnet can easily be used as a diagnostic test to tell the difference between platinum and steel. Lead is a lot softer than platinum and will smear when subjected to a knife blade.

Many of these tests were and are used when prospecting. They are fast, easy ways to determine whether or not some economically valuable mineral might be present in a placer or in an outcrop without having to wait for a lab to return a chemical analysis. Fast definitive field tests were best because for many studies I only had one field season to evaluate an area. If it could be determined there was a valuable material in an outcrop without waiting six months for a chemical analysis, then I could follow-up the investigations with more work, like soil sampling.

Understanding geology can have a big influence on prospecting. When the BOM was studying the Colville Mining District in the National Petroleum Reserve Alaska (NPRA), previous studies had noted lead-zinc deposits along the Brooks Range, starting at around the Red Dog Mine on the west to the Dalton Highway on the east. Joe Kurtak of the BOM noted a lot of these deposits were at the base of thrust faults. A thrust fault is a fault with a shallow dip where the overlying rock is older than the underlying rock because it has been pushed over the underlying rock. Because of this revelation, Joe's crew would fly along the Brooks Range until they noticed a thrust fault, then they would stop and walk across the fault looking for mineral deposits. It was a very successful strategy and one that resulted in the identification of a lot of new deposits.

Evaluating a property in the Juneau Mining District

Chapter 15

Property Evaluation

Nathan Rathbun and I were looking for an abandoned underground mine on the north side of Shoup Bay near Valdez. The helicopter let us off in an open area and we plunged into the heavy brush. We had an idea where the underground working should be, but couldn't see more than a couple of feet in front of us. After crossing the slope for an hour, we decided we were probably above the workings, so we started downhill. We had only gone a couple of hundred feet when Nathan disappeared. He yelled, "I think I found it." He had fallen about 6 feet into the trench in front of the mouth of the working. Luckily the brush was thick enough to break his fall. The next thing he yelled was, "Bees!" He jumped off the end of the mine dump and ducked as a swarm of bees passed over his head. We waited a while for the bees to clear out of the entrance, then started evaluating the property. This would be one more evaluation out of 1,000s I conducted in Alaska.

Evaluating mineral properties should be simple and easy. All that has to be done is find, map and sample the property; however, it's a little more complicated. First, the property has to be located. If a property is discovered during exploration then it's a no-brainer. Voila, the property is located. However, a property someone noted either on a map or in the literature can be problematic. In some instances, there is a good location on a map and it actually exists on the ground. In other instances, it is like prospecting for an unknown deposit. Sometimes the only mention of the property in the literature is someone staked a claim somewhere in the area. One property my crew looked for just had a claim notice stating someone had staked a claim on the west side of an island in Prince William Sound. After walking the whole west side of the island a number of times, no one ever found anything that would have justified staking a claim. But since the island is so heavily vegetated, we probably wouldn't have noticed any kind of disturbance unless we stumbled or fallen into a prospect pit.

Besides non-explicit property locations, time is always a factor. When I was working in Prince William Sound, I was trying to locate and evaluate over 600 identified mineral properties. There were less than 180 days over 3 years to accomplish this work. The additional time spent following-up geochemical anomalies and prospecting for unknown deposits meant the crew had to find, map and sample over 3 properties per day every day. And, for those properties with over a half mile of workings, mapping and sampling might take all day. Even with two 2-person field teams, there wasn't enough time to do justice to every property, if and when a property was found.

If a geologist is lucky, a property location is obvious. The obvious ones have holes where miners dug into the side of a mountain and dumped their spoils down the slope. The spoils create a mine dump or a pile of rocks and dirt in the form of a wedge on a usually uniform mountain slope. Many times these mine dumps are located in the most remote and seemingly inaccessible places. When dumps were found in these bizarre places, I could only surmise the "old timers" were superhuman or the work was done by space aliens. I also surmised some of these "old timers" must have thought the best deposits can't be easy to find and must be in the worst places; therefore, the "old timers" spent all of their time and money looking for the "mother lode" in places avoided by most God-fearing people.

Sometimes, the properties are just so obvious it's a wonder the area isn't crawling with geologists. While walking up a creek in the Koyukuk Mining District Joe Kurtak, Mark Meyer and I rediscovered a porphyry copper deposit previously claimed in the past by a mining company. A porphyry copper deposit is one where there are many veinlets of chalcopyrite in an intrusive rock. Usually, these deposits are low grade, but big. Porphyry copper deposits are commonly mined in Utah (Bingham Canyon), Arizona (Morenci, Ray, Bisbee) and New Mexico (Chino). While walking up a creek and looking at copper veinlets in the canyon walls, I imagined how the old-time prospectors must have felt when they discovered the copper deposits in the southwest U.S. in the 1800s. One of the wonders of evaluating mineral properties in Alaska is a geologist can see many mineral deposits in their virgin, untouched form.

Once a property has been found, the real fun begins. The property has to be mapped and sampled. If it is warranted a property will also be drilled and possibly some geophysical surveys conducted.

The author mapping an
underground property

Chapter 16

Mapping

We stretch the tape and set it on the floor. I take out my Brunton compass and shoot a bearing on my field partner's headlamp 100 feet away. The light from the headlamp is the only thing visible in that stygian tunnel. I put away the compass and take out my notebook to record the compass bearing. I then slowly walk up to my partner while recording the width and geology of the tunnel in my notebook. I asked my partner if he had seen any copper minerals. For the umpteenth time, his answer was "Nope" as he grabbed his end of the tape and proceeded further into the depths of Copper Mountain in Prince William Sound. We were almost a half mile into the mountain at the Reynolds-Alaska Mine and the only thing visible was the volcanic rock as black as the void in which we were mapping. The tape stopped and my partner said, "We've reached the end and guess what, there's finally ore." It looked like the miners found a mass of copper sulfide and stopped mining. Curious. Why would a company expend so much money to develop a property, finally find what they told their investors they were looking for, then pack up and walk away? Well, maybe the whole point of the exercise was to mine their investors, which sometimes is a lot more lucrative than actually trying to profitably mine something. But, it didn't matter because our job was to document what was present and not try to speculate about the past. I said to my partner, "Let's take our samples and get while the gettin' is good." We took our channel samples, gathered up the sample bags, put them in our packs and slogged a half mile until we were finally out into the fresh air. Whew! Another property found, mapped and sampled. Only a couple of hundred more to go and the fieldwork portion of the project will be over for the season.

Mapping is where geologists use their years of education. Much of the work of a field geologist can be done by anyone who is physically fit because it just involves walking, carrying a backpack and hitting stuff with a hammer. However, mapping requires a lot of school-acquired knowledge. This includes being able to identify rocks and minerals in the field, map reading, and surveying.

There are various levels or scales of mapping. Most academic or survey geologists map at a small scale versus a large scale. The words to describe scales are kind of confusing. Small-scale means objects depicted on a map will be small in size. Large scale means objects depicted on a map will be large in size. When I was taught mapping in college, worked for the USGS and mapped my thesis area, I mapped at a small scale. An inch on the maps I made represented 2,000 feet on the ground. If an object on the ground was less than about 100 feet wide, it was depicted on the map as a pencil line. It was as if I was making a map of a city, which only showed the streets. In contrast, mineral property maps I made were drafted at large scales where an inch on the map might represent anywhere from 10 to 50 feet on the ground. These maps were like making as-built surveys of buildings within a city. This level of detail is needed in order to describe a mineral property. Ideally one would map all of the details at a deposit, such as the geology; the location of the mineralized material; the dimensions of the deposit; any workings present on the surface and underground; the rock structure, like faults, folds, strike and dip; and mine structures, like head frames and mills. These maps were used to determine the size of a deposit, which was usually calculated in the office after the fieldwork was completed. A detailed map showing the width and length of the mineral deposit was therefore very important.

The detailed mapping exercises I had in college were rudimentary. I was fortunate while working for a company in Colorado that they required me to map in detail using a compass and tape. This detailed mapping skill was later honed when I worked for BLM in Alaska. BLM sent me to their National Training Center, which taught very detailed mapping of surface and underground mine sites. After I left Alaska in 1995, I worked for this Training Center teaching new geologists the mapping skills I had learned 16 years previously.

Detailed mapping skills proved invaluable in Alaska where I had to meticulously map mineral deposits and mine workings for WGM, BLM and BOM. This was especially true when I mapped a lead-zinc deposit WGM had found north of the Yukon River. After spending all day mapping the geology around the deposit, I came back with my map and showed it to my boss. This was the boss who had told me since he hired me, he could fire me. He questioned my mapping and asked me if it was right. He shook my confidence, but I assured him the map was correct. The map depicted folded rocks with the lead-zinc minerals in the middle of the fold. I didn't think he had much faith in a 24-year old's mapping skills because he said he'd take my map and check to make sure I hadn't screwed up. I was worried all day while working on another property if he would come back, tell me how I messed up and either I would forever be an idiot in his eyes or be looking for another job. When we both returned to camp he admitted the map was right. Whew.

I mostly used these detailed mapping skills working in southern Alaska, where there are a lot of underground workings. It seems most underground workings are near the coast, either in Prince William Sound, Cook Inlet or Southeast Alaska, where there are 1,000s of workings. In the interior, many of the deposits I evaluated had surface exposure and could be mapped in less detail.

Not everyone thinks working in an underground mine is fun, but I do. A mine is pitch black. If everyone turns off their headlamps, no one can see their hand in front of their face. I always carried at least two backup sources of light, just in case, the primary light source went out. The workings are usually less than 6 feet wide and from 6 to 7 feet tall. Tall people, like me, have a tendency to bang their head on the ceiling. They can be wet and cold. I mapped many workings in my rain gear where the water dripping off the roof would run down my back. The workings can be claustrophobic. Some of the workings I mapped were thousands of feet long with only one way in or out. I always ask myself when I am in one of these mines, "What do I do if something happens?"

This question was spinning around my head during one mine evaluation in Prince William Sound. Bob Hoekzema, Gary Sherman, Nathan Rathbun and I went to an abandoned gold mine located in College Fiord. It was nestled under a glacier hanging from the mountainside. The mine consisted of a crosscut into the mountain and then drifts along a quartz vein. A crosscut is an underground working that cuts across barren rock to access a mineral deposit. A drift is an underground working along a mineral deposit, like a vein. A couple of hundred feet into the crosscut, there was a tee. In the tee was a quartz vein. The old miners had mined the quartz vein about 100 feet on both sides of the crosscut. At the end of one of the workings, the miners had left cases of dynamite stacked from floor to ceiling, filling up the whole space. We didn't approach the dynamite because there was water on the floor at the foot of the cases and it looked like an oily sheen was on the water. Nitroglycerin will leach out of dynamite and make anything it soaks into explosive. It is also oily and will float on water. Therefore, we were wary of getting anywhere near it. When our crew finished mapping the workings Bob, Gary and Nathan went outside and left me to sample the quartz vein in front of the dynamite. Being alone and pounding on rock with a hammer and chisel while standing next to tens of cases of dynamite was unsettling, to say the least. I was also thinking this mine was close to the epicenter of the massive Alaskan Good Friday earthquake of 1964 and what would happen if there was another earthquake while I was sampling. Needless to say, I took the samples as fast as possible and promptly exited the mine. This site was reported to the Forest Service as an obvious hazard to life and limb. In the early 1990s, the BOM with help from the Forest Service and the military disposed of the hazard by blowing up the dynamite. The BOM researcher said the explosion looked like a cannon shot coming out of the mine opening.

After crudely mapping a property in the field in either

my notebook or grid paper, I spent time, meticulously drafting the results onto 8 by 11-inch mylar sheets in the camp office. I would take extra time making sure that what I saw on the ground was depicted on these sheets. There were many times where I looked at a property and wondered if I was in the right place because the map that I had found in the literature was different than what I saw on the ground. After assuring myself that I was in the right place, I could only assume that the original mapper had transposed their bearings when they drafted their maps. I never wanted to make that mistake on the maps that I was drafting; therefore, I always double-checked my field map with what the working looked like before I left a property. After I finished drafting the map in the office, the mylar sheets were hole punched and put into notebooks with the daily work reports. These maps were redrafted in the office by a draftsman for inclusion in the final reports.

Mike Balen and USBM employee using a backhoe and high-banker to sample a placer property in the White Mountains

Chapter 17

Sampling

"Man, you are the best digger I've ever seen," I said to the seasonal firefighter helping me and the BLM District Geologist sample a placer gold deposit. The District Geologist chimed in he'd never seen anyone dig such a deep hole in so little time. We were evaluating a bench gravel deposit in Chicken, Alaska and were able to persuade a 19-year old kid from the BLM firefighting crew to help us. He didn't take much persuading since it had been a wet summer and therefore a very slow fire season. Fortunately for us, it seemed all we had to do was heap on the praise and the gravel flew out of the hole. It wasn't as if the District Geologist and I weren't busy. We were holding the sample bags open and as they got filled with 100 pounds of gravel, we would lug them over to the truck. Eventually, we would have to process the gravel through a device in order to extract the gold, so we weren't just being jerks, we were pacing ourselves. As the firefighter finally hit bedrock and finished the hole, I asked the District Geologist how many samples we needed to take to properly evaluate the bench gravel deposit. He said he thought five more samples on each side of the bench gravel would be enough to figure out if gold was present in paying quantities. Unfortunately, even a 19-year old who thinks of himself as a human backhoe has limitations, so to get all of the samples taken, the District Geologist and I had to eventually grab the business end of the shovel and get digging.

This mineral property in Chicken Creek would be just one of countless occasions where I had to sample a mineral deposit to determine if and how much "good stuff" was present. In order to properly evaluate a deposit, a person has to take samples in an appropriate manner and at appropriate intervals. These intervals are dictated by the geology of the deposit, the way the valuable minerals in the deposit might be distributed, the available time a person has to take samples, costs of taking a sample and how important the samples are to the study. Much sampling involves brute force and carrying equipment, supplies and in the end full sample bags.

The easiest kind of sample to take is a grab sample. To take a grab sample a person just smacks a piece of rock off of an outcrop with their hammer or picks it up from the ground. A grab sample is usually the best-looking rock found at a mineral property or prospect. One of the disparaging names for us at the BOM was the Bureau of High Graders because we took grab samples. In our defense, people had to realize a grab sample can never be used to determine the viability of a deposit. It was only taken because there was usually not enough time to adequately sample a deposit (remember 3 properties per day). Because of the lack of time needed to take detailed samples, there was a good likelihood we wouldn't take a sample of "good stuff" that was present at a prospect. So, by taking a grab sample we could at least get some idea whether or not there might be anything good on a property.

I once mapped and sampled a small underground working on a quartz vein along the edge of Columbia Glacier. I took a couple of chip channel samples across the vein inside the working and then proceeded to look through the rocks lying on the mine dump. While sitting on the dump picking up pieces of quartz, the helicopter pilot asked if gold was ever found on the dumps. As he asked me this question I picked up a piece of quartz that contained gold and showed it to him. Then I picked up another piece and another piece. It looked like the "old timer" who had made the hole in the ground had bagged up his best samples and left the bag on the mine dump. The bag had deteriorated, leaving the gold-bearing quartz in a pile. In contrast, the samples taken from the quartz vein did not have any gold. This experience just emphasized this property might have a higher potential to contain gold than what I could find in outcrop.

The importance of a property evaluation also has to be considered before a grab sample is taken. I never took grab samples when doing regulatory work. Regulatory sampling has to be defensible in court. The last thing a person wants to do after systematically sampling a property and not finding enough mineralized material to make it a valuable mine is to try and argue a property is no good if they had taken a grab sample with a high value. For the BOM, we never had to defend our sampling programs in court, except the court of geologist's opinion; therefore, grab samples were OK.

The most common samples I took when evaluating a mineral property were channel or chip channel samples. Ideally, a geologist takes a channel sample across the width of the exposed deposit. When sampling an outcrop of gravel it was important to try to cut a channel about the width of the head of a shovel from the top of the gravel bed to bedrock. When an outcrop of rock was sampled, the ideal rock channel would be 2 inches deep and 4 inches wide across the rock exposure. When completed a 2 x 4 piece of wood should fit in the channel. Even though a channel sample is ideal, it was rarely taken because of a number of factors. The main factor is that cutting a channel sample is usually very difficult. For placer samples, the gravel layer might be too thick to be able to reach bedrock with hand digging. For hard rock deposits, many times the rock is too hard to cut a channel using a hammer and chisel. One time I tried to sample a quartz vein where there was a beautiful 2 inch by 4-inch channel someone had previously cut. I used a 2-pound hammer and a chisel and pounded on the rock for half an hour. All I could do was break off a couple of chips of rock that happened to go flying in my face and down the slope. Therefore, if the gravel is too thick or the rock is too hard then it's almost impossible to take a channel sample without using some kind of powered equipment, which was usually impractical in the Alaska bush. Another factor is why do I need to take a channel sample? If I am just trying to determine if a deposit has a low, medium or high potential, then a channel sample is overkill. I only took channel samples when I did regulatory work and had to defend the results of the sampling program in court, or if the mineralized outcrop was soft enough to easily cut a channel. If it was for important enough work, I could always get the necessary equipment needed to take those kinds of samples.

The most common sampling I did was chip channel sampling. I would try to chip pieces of rock in a straight line over the width of the mineralized outcrop. I would lay a tarp on the ground and let the chips fall on the tarp. The chips would be put into a sample bag. It was important to take chip channel samples of mineralized material and a separate sample of barren or wall rock material. By separating the samples, the true width of mineralized material can be calculated and it could be determined whether or not the wall rock material had any value. There are different mining methods employed that can selectively mine the mineralized material, so it's important to know the extent of mineralization or the "good stuff" prior to developing a mine plan.

It is also important to take a wide enough sample. Mark Meyer discovered a mineralized gold vein in the Valdez area. The vein is about 6 feet wide, but the gold is concentrated in only 3 inches of the vein, near the edge. A sample taken from the 3-inch wide gold-bearing portion of the vein assayed over 100 ounces per ton. However, as they say, "There ain't no 3-inch miners," so the entire width of the vein was also sampled, which resulted in much lower values. Six feet is commonly considered a minimum mining width.

People have to be careful about using the results of any sampling programs without independently doing their own work. I met some underground miners in the Valdez Creek area. They were mining a quartz vein based on a sample taken from the face of an underground working by a State geologist. They were from the lower 48 and had brought all the underground mining equipment up with them. They were in their second year of work and had not found any gold. They were complaining about the geologist who took the sample. I thought, "Wouldn't you go in and independently verify the results before investing all of your time and money?" The answer I guess was, "No." I got the distinct impression the miners thought it was a good excuse to get away from home, come to Alaska and play with equipment.

I also mapped and sampled a tin property near Lake Sithylemenkat when I worked for WGM. During one of my traverses across the property, I found a black rock lying on the ground. It really stuck out against the tan colored rocks in the area. I bagged it up and the subsequent analyses returned a high lead value. I was asked about the rock by WGM geologists for years afterward, but couldn't really add anything more than it was just lying there. It is common knowledge some geologists after examining a property might leave some samples collected from other deposits. This is done to confuse any rival companies interested in the same properties. It just means ya gotta be careful out there and determine if a rock fits with the other rocks at a deposit.

Besides taking grab, chip or channel samples from a property, I might take a large or bulk sample if it was warranted. Bulk samples were taken in order to figure out the characteristics of a deposit. These characteristics, like grain size and mineralogy, are critical in trying to determine the economics of mining a deposit. These characteristics are needed to design a plant to separate the good stuff (i.e., minerals a miner gets paid for) and the bad stuff (i.e., the minerals miners throw away). Bulk samples are easy to take but labor intensive. For, lode deposits, I would bag up about 2,000 pounds of rock and ship it to the BOM labs in places like Salt Lake City, UT or Reno, NV. These labs would handle the rock like any mining operation, (i.e., crushing, grinding and processing).

Bulk samples were also taken to characterize a placer deposit. I and whoever was helping me, would shovel about 1 cubic yard (3,000 pounds) of gravel onto a screen with 1-inch squares, weigh the oversize and then bag the gravel that passed through the screen. This bagged gravel would be taken to the warehouse in Anchorage. Nathan Rathbun would sort the gravel using various sized screens. The material from each screen size would be panned and any gold recovered would be removed and weighed. This information could be used by someone to design a very efficient processing plant.

Bulk placer samples were also taken and processed in place. This was done by shoveling or using a piece of equipment like a backhoe to dig out thousands of pounds of gravel and put them into a machine to separate out the gold. Gold savers or high-bankers were commonly used for this purpose. A gold saver is a machine mounted on a trailer for transport. The machine has a hopper where the gravel is fed. The gravel is washed from the hopper into a rotating drum or trommel with slots. The over-sized material moves down the rotating drum because of gravity and exits the drum via a chute on the back of the machine. The undersized material is washed through the slots and into a sluice box mounted underneath the trommel. The machine is powered by a gasoline engine. Gold is caught in the sluice box. This is a highly efficient machine for gold recovery and mimics how a miner processes gold-bearing gravels at a placer deposit. I used this machine when I worked for BLM to evaluate a placer deposit along Chicken Creek and also at Kantishna.

Sometimes a high-banker was used, which is kind of a poor man's gold saver. A high-banker is what I call a two-piece contraption with a screen sloping one way, which is mounted over a sluice box sloping in the opposite direction. The screen has a spray bar attached to a hose with a water pump. Gravel is dumped onto the screen, with the oversize falling off the back and the undersized going through the sluice box. Water volume can be controlled in order to get maximum recovery. This machine is light and can be set up anywhere there is a water source within reach of a hose. I used it in Goodnews Bay to sample beaches, along creeks throughout Alaska to sample bench gravel deposits and at Nome Creek during the White Mountain study.

Many placer properties are mined using suction dredges. These properties are usually streams or rivers where gold has concentrated on the bottom and have little gravel on their banks. To properly evaluate these properties, the BOM used a suction dredge. A suction dredge is simply a water pump and sluice box sitting on floats. A hose attached to the pump is used to vacuum the gold and gravel off the bottom of a creek or river. Dredges come in various sizes, some using hoses 2 inches in diameter run with lawnmower engines to hoses 10 inches in diameter powered by car engines. The material vacuumed off the bottom of the drainage is put through the sluice box. Any gold in the gravel is caught in the riffles and the lighter material is expunged out the back. The size of the dredge used on a creek or river depends on the size of the water body and how much money the miner has to spend. For shallow drainages, a small, 2-inch to medium-sized, 4-inch dredge is OK. The BOM used a 4-inch dredge to sample a drainage. The sampler usually knelt in the creek bed and vacuumed up the gravel, while his partner, who was usually me, threw rocks bigger than the hose away from the sampling area. A sample taken using a dredge would indicate whether or not there was gold in a drainage and how much gold might be recovered in a set amount of time.

If the mineralized outcrop at a deposit is not very well exposed, a basic method used is soil sampling. A sample of soil will indicate what might be hidden beneath the surface. If a buried deposit was suspected, then a soil grid would be run over the deposit. A soil sampling grid consists of a series of parallel lines, which are so many feet apart. A sampler walks along the lines and takes a sample of soil at certain intervals along the line. Soil consists usually of three horizons, A, B and C. The A horizon is at the top. The B horizon is in the middle and usually contains fine-grained material. The B horizon is what most people think of as soil and where farmers plant their crops. The C horizon is broken up country rock above the hard bedrock. In Alaska, in mountainous terrain, there are usually only the A and C horizons. There are few well-defined B horizons in these terrains. Also, much of the ground below the A horizon is frozen, which makes getting a good sample difficult. To take a soil sample in Alaska, I would usually just use my rock pick and dig a hole. I would then reach my hand into the cold dirt, grub out any dirt and put it in a paper sleeve.

The first mineral discovery I ever made in Alaska occurred while taking soil samples. WGM had found a little lead-zinc deposit along one of the drainages to the Yukon River. The project leader wanted to know the extent of the deposit, so he sent me to take samples every 50 feet across a slope in a drainage east of the deposit. After 4 hours of kneeling and scooping dirt into paper sacks, I sat down to eat lunch. While eating, I noted a slough area below me. It looked interesting but I was reluctant to climb down 50 feet, knowing I'd have to climb back up to resume soil sampling. However, after lunch and a rest, I was refreshed enough to say, "Heck I'll go take a look." What I found was a massive sphalerite (zinc sulfide) outcrop about 3 feet thick. I don't think the discovery ever amounted to a major deposit, but it was nice to find something no one else had ever seen. I took a couple of samples and climbed back up and finished the grid.

My boss sent me back to the outcrop the next day to take a chip channel sample across the outcrop. While sampling I looked up and saw a big black furry face staring back at me from above the outcrop. My heart started racing as did my feet. I ran to the nearest tree and kept watch until the helicopter came to pick me up. This was my first year in Alaska and I wasn't carrying a firearm. At the time my only defense against a bear attack was my trusty rock pick or trying to climb one of the spindly black spruces. Neither were good choices and luckily, I guess, I didn't look too tasty. I must have gotten too grizzled-looking after a month in the field or the thought of eating a hunk of meat slathered in bug dope didn't seem so appealing to this bear's discerning palate. Whatever the case, I was glad when the bear resumed its wanderings.

Even though the size of the outcrop at my discovery is small, I can compare it to the small outcrops I have visited at some of the biggest deposits in Alaska. The lead-zinc bearing outcrop in Red Dog Creek at the Red Dog mine was only about 4 feet thick. The discovery outcrop of the Arctic deposit, which is a huge copper deposit in northwest Alaska was about 3 feet thick. It took years of drilling to eventually reveal the large sizes of these deposits.

The big question that doesn't really have an answer is, "How many samples must one take and at what intervals in order to adequately define a mineral deposit?" I asked an oil geologist how many holes did she have to drill to adequately define an oil field. She told me all she needed was one hole. Ah, if it was that easy with every other kind of mineral deposit. Mineral deposits are uniformly non-uniform. Mining people can usually answer the question after they have mined out a deposit. Geologists try to do the best job of sampling a deposit that time and money will allow, but they know there are numerous examples where mines have not been successful because of inadequate sampling. Therefore, the answer to the question of how many samples and at what intervals always rests on the questions of how important is the study the geologist is conducting and what kind of deposit is being sampled.

Regulatory work requires very detailed sampling because the results possibly have to be defended in court. Much of the work for the BOM was to determine if a deposit had a high, medium or low mineral potential. This kind of determination could be done with minimal sampling. One study the BOM conducted, in Kantishna, was done for the National Park Service. The results were used to determine the amount of compensation claim owners would receive from the government when the government confiscated their mining claims. The BOM conducted extensive sampling to make the determinations.

The types of deposits sampled were a big determining factor in the number and intervals of sampling. Gold deposits are invariably variable. At one property, I took 2 samples 6 inches apart from a quartz vein. One sample contained over 1 ounce of gold per ton, the other sample had no gold. When I got the results back I had to ask myself how many samples were needed to adequately sample a gold-bearing vein? The answer was, "Too many" for the kind of study that was being conducted. Needless to say, in the end, a geologist must rely on their professional judgment to answer those questions.

A final note about sampling is the use of sample cards. A sample card is used to document where a sample is taken, the date, a description, the kind of sample taken and possibly what economic elements are suspected. Every entity that takes samples either buys off-the-shelf cards or in most cases develops their own sample cards and has the cards printed. Every card has either a unique number on the top of the card or a blank space where the sample number is written. Every card has a small rectangular, perforated area, which contains a duplicate of the sample number. The perforated rectangle is detached from the card and put into the sample bag. The sample number is either written on the sample bag either on the bag itself with a magic marker or on a tag attached to the bag. Some geologists prefer to use blank metal tags attached to wires. The wires can be tied to the bag and the sample number can be embossed on the metal. How so ever anyone labels sample bags, it was extremely important that there are multiple ways to identify the sample in case something happens to obliterate one of the sample numbers. The configuration of sample cards is another one of those conversations geologists have while consuming adult beverages. I contend that there is no perfect sample card; therefore, the quest for a perfect card gives geologists a perfect excuse to consume vast quantities of liquid refreshment while debating the elusive ideal.

Cable tool drill

Chapter 18

Drilling

"Hey, come here, look at this," I said to the driller and his helper as I stood next to a pile of brush with a moose leg sticking out of it. "What the heck? This wasn't here yesterday. Who or what did this?" said the driller's helper, who hadn't been in Alaska very long. The driller and I explained that a moose must have wandered near our drill site last night and a hungry bear decided it looked good enough to eat. We also explained that when a bear kills a moose, the bear eats what it wants and buries the rest. "Do you think the bear is still around?" said the driller's helper. We told him there was a likelihood the bear was somewhere in the bushes sleeping it off and we should just get back to work, keep our eyes open and I'd keep the shotgun close at hand. I had found the bear's food cache during my morning walk-about. I, as the geologist monitoring the placer drilling in Denali National Park during the BOM's Kantishna Study, always did a morning sweep of the area around the drill rig just to make sure there were no grizzly bears lurking in the bushes. I also didn't have anything to do until the drill started producing samples, so I just tried to be useful and not get in the way. The drillers and I agreed the best thing to do was to get the drilling done as fast as possible and move to a site that wasn't next to a fresh bear kill.

Most geologists will at some time in their careers be involved with a drilling project because a geologist has few options to choose from if they want to determine what is below the surface. One option is to dig a trench. Most mineral locations are pretty remote in Alaska, which means there is little chance to bring trenching equipment, like a backhoe to a site. Geologists therefore usually grab a shovel and start digging or in the past used dynamite to blast a trench.

When I first came to Alaska in the 1970s, buying dynamite was pretty easy; therefore, dynamite trenching was a common practice. One of my first assignments with the BOM was to go buy a case of dynamite, load it in a chartered plane, fly to Cordova and deliver it to the road department. A BOM seasonal employee who was in a remote camp had borrowed a case of dynamite from the road department, so he could do some trenching. Before he could get a trench blasted, a bear got into the case and destroyed the dynamite. Luckily I was stupid enough to only be a little nervous flying in a small plane over the ocean with a case of dynamite next to me. The government cracked down on the purchase of dynamite, so unless a certified blaster is available, this type of trenching is less common today.

A geologist will only dig a trench so deep, which usually isn't very deep. Therefore, the only real option to determine what is below the surface is to drill. Surface exposure indicates the length and possible width of a deposit and even grade, but not the depth of the deposit. There is a rule of thumb used to determine depth if the only thing a geologist has to go by is the surface expression of the deposit. The rule of thumb is a deposit's depth is half the distance of its length. Even though I've calculated depths many times using this method and it will stand up to scrutiny, because it is a standardized rule of thumb, it's still not a very good measure. For example, if a vein is only exposed for 10 feet, then by using the rule of thumb, it only goes 5 feet into the earth.

Many times surface samples of a mineralized outcrop do not depict the true grade of the deposit because the deposit has usually been exposed to weather for an unknown number of years. Weathering can either increase or decrease the grade of a deposit. I've looked at many production records for gold deposits. Most of the records showed when gold veins were found, the grade mined from the surface was over one ounce of gold per ton. However, as the miners dug deeper into the deposit the grade dropped to below 1/2 ounce per ton. Weathering will also leach metals from many sulfide deposits leaving behind a surface largely devoid of metals.

Because of the weathering factor, a fresh sample is preferable to determine the true grade of a deposit. The best way to get a fresh sample is either by drilling or driving an underground working. In the old days, most prospectors (aka the old timers) realized they wouldn't be able to tell whether or not a deposit was economic to mine unless they looked below the surface. So they either sunk a shaft (a vertical hole in the ground) or drove an adit (a horizontal hole in the ground). This work was a little easier to accomplish back then because the old timers only needed their camping gear, some dynamite, hand steel, a hammer and possibly a wheel-barrow. A prospector would use hand steel, which is a round piece of iron with a star-shaped bit on the end. The old timer would pound the steel into the ground with a hammer, making a hole. Once enough holes were made, usually in the shape of a curved doorway, approximately 6 feet wide and 7 feet tall, the holes were loaded with dynamite. The dynamite was set off, breaking the rock. The broken rock was hauled away in a wheel-barrow or just thrown down the slope.

Making a large hole in the ground was preferable for these early prospectors because it was cheaper than buying or hiring mechanized equipment and hauling it out to their prospect. This was especially true for some of the remote prospects I have examined. As the years wore on and the prospectors who knew how to drive tunnels by hand got too old and died off, they were replaced with machines (i.e., drilling machines). Therefore big holes in the ground were replaced with small holes.

There are several types of drilling rigs. I have been involved with many of them throughout my career. However, in Alaska, I was only associated with a hand-held drill, a "Winkie" drill and a cable tool drill. The BOM had a hand-held drill capable of making a hole about 3 feet long. It consisted of an engine, about the size of a lawnmower engine, attached to a bracket with hand holds. A drill rod was at the end of the engine. The drill rod was hollow, so as the hole was drilled, the rock would be pushed up into the drill rod. When the operator was done drilling, there would be a core sample about 1 inch in diameter. I used this in areas where the surface of the rock was smooth and wanted to get some idea what was beneath the surface.

Another drill was the Winkie drill. A Winkie drill is a portable drill. It is a larger version of the hand-held drill. It is capable of drilling up to 500 feet. It is essentially light-weight (less than 200 pounds). Many times the drill is mounted on a tripod. The drill has an engine, transmission and water hoses. The drill is powered by hand, which means the operator applies pressure on the drill rod by hand. The drilled rock is pushed up into the drill rod. Every few feet the rock is extracted from the drill rod and put into a box which is labeled with the depth where the core was taken. The core is about one inch in diameter. These were great machines for helicopter-supported operations. Winkie drills were used during the BOM study of the Kantishna Mining District and during the BOM strategic and critical minerals studies. The geologist's job during this type of drilling is to describe the type of rock that comprises the drill core, split the core down the middle and send one half in for chemical analysis.

The third type of drilling I was associated with in Alaska was cable tool drilling. A cable tool drill is used in placer deposit evaluations and is very common in the lower 48 for drilling water wells. According to Wikipedia, it was invented by the Chinese over 2000 years ago and has been used to make holes ever since. A cable tool drill is a large contraption. It is usually mounted on a truck, so road access is needed. The drill consists of an engine, an arm about 20 feet off the ground, a steel cable, a large rod of solid iron and a bailer. The large rod is attached to a cable strung on a pulley on the tip of the arm. The engine is used to raise and drop the solid piece of iron. The iron rod, which is about 10 feet long has a cutting tip at the end. The tip breaks the material when the bar is dropped. As the iron rod is raised up, the rod rotates slightly so the next hit isn't in the exact spot as the first hit and therefore it causes more breakage. The iron rod is raised and dropped, raised and dropped until it can no longer break anything because the rod is hitting broken material. Then a bailer is attached to the cable.

The bailer is a hollow tube with a trap door at the end. It is lowered into the hole, the broken material goes into the end and as the bailer is raised the trap door closes. The material is brought up to the surface, emptied into a trough and the process is repeated until no more broken rock is extracted from the hole. If the drilling is in frozen ground, then the hole will remain open. However, if the ground is thawed, then a drill pipe approximately 8 inches in diameter must be pounded into the hole to keep it from collapsing. The average placer cable tool drill can drill about 20 feet per day.

For the Kantishna study in 1983, a cable tool drill was contracted to determine how much gold might be in the gravels on the benches above the creeks. Bench gravels are those gravels deposited in the past by a stream. The bench is made when the stream for whatever reason cuts down, sometimes tens of feet, leaving the gravels behind. The drill arrived at the beginning of September, which is a tricky time of year, weather-wise. There are days of sun and other days of snow. The main constant is dealing with frozen water in the morning. I was assigned to oversee the drilling because many people on the field crew had better things to do, like moose and caribou hunting. People, therefore, were abandoning camp at a furious pace. I felt I was in a game of musical chairs, where the prize was a get out of camp free card and I was the slowest one who happened to not find a chair at the end of the game. Therefore, since I lost the game, I was in charge of watching the drill and dealing with the samples. During the drilling program, I independently verified the figure of 20 feet a day because there is not much to do when watching a cable tool rig drill besides figuring out why the drilling is taking so long. The only fun parts of the job were panning the samples extracted during drilling and discovering interesting things like buried moose carcasses. Luckily at the dead moose location, we were able to get the drilling done quickly and move to the next location before dark. Fortunately for me, we were supposed to drill until the end of September, but the money ran out after a couple of weeks and as a result, I finally got my get out of camp free card.

USBM geologist outfitted to run a geophysical survey in Kantishna

Chapter 19

Geophysics

"OK, the lines are laid out, the metal plate is on the ground, the sensor cord is attached to the sledgehammer and recording device. It's time to make some seismic waves," I said to my co-worker. The co-worker swung the sledgehammer high over his head and down onto the metal plate. Wham! Whoops, we missed a step. We forgot to secure the sensor cord to the sledgehammer's handle and it fell between the head of the hammer and the steel plate, which severed the cord. We looked at the end of the cord and there were 5 red wires. Being scientists, we ran through all of the ways a person could splice 5 wires together and came up with over 3,000. Nope, reattaching the wires in the correct configuration wasn't happening. Better get this thing back to the BLM shop in Fairbanks. Our first attempt at using geophysics to determine the depth of gravels at Chicken, Alaska was a failure. So much for trying to be geophysicists.

Geophysicists or geowizards (my term) use electric-powered machines to determine what might be underground. My name for these machines is gee-whiz machines. There are various types, like magnetometers, scintillometers, ground penetrating radar, induced polarization, and seismic units. Usually, these geophysical studies are carried out by the geowizards, but field geologists can do some of the basic work.

To get meaningful results, the machines must record data at specific points on a grid. Field geologists are very good at laying out those points because of their mapping skills. Joe Kurtak and I got assigned this duty during the beginning of the Kantishna study when the project geophysicist wasn't able to get any of his non-geologist helpers to reliably survey in his points. Joe and I filled in for about a week while we taught the non-geologist helpers how to survey. This type of surveying just involves stretching out a tape, reading a compass and pounding in a stake at the appropriate point. Joe and I decided after laying out thousands of feet of line and hundreds of survey points, the work was more suited for the newly trained helpers. Therefore, after training the helpers we handed off the tape and stakes and moved onto jobs more suited to our education and experience.

Field geologists are also used by geophysicists to carry equipment. Because a field geologist is used to humping a pack around all day, geowizards are only too happy to foist their machinery onto a strong back while they go off and record the collected data.

Field geologists with a little training can do simple geophysics work and interpretation, like the work associated with seismic, magnetometer or radiation studies. A seismic machine records data when energy is transmitted into the earth. The energy can be transmitted by setting off an explosive, by using vibration or by using a sledgehammer with a wire attached to it. Oil companies that look 1,000s of feet into the earth use explosives and/or trucks that vibrate the earth. For some studies, a person can use a small seismic unit to look 10s of feet into the ground by hitting a piece of steel on the ground with a sledgehammer. The unit should record where the seismic waves pass through the overlying gravel material and into the underlying bedrock because of the difference in their densities. This is what I tried unsuccessfully to do at Chicken. I took this experience with me when Bob Hoekzema, Nathan Rathbun, Gary Sherman and I used the same kind of machine to evaluate the bench placers on the Kenai Peninsula. As they say, experiences are gained by surviving the mishaps.

Field geologists do some geophysical studies using some simple gee-whiz machines. They take a hand-held magnetometer or scintillometer and walk over an area recording readings every so often. A magnetometer measures either the magnetization of a magnetic material or the direction, strength or relative change of a magnetic field. Some buried mineral deposits can be delineated using magnetometers because they have different magnetic signatures relative to the surrounding rock.

A scintillometer, which most people would incorrectly call a Geiger counter measures radioactivity. A geologist uses a scintillometer to look for radioactive material such as uranium. Some geologists know more about geophysics than others, but I'm not one of them. I have used magnetometers and scintillometers at various times, but only to tell me whether or not one reading was better than another. I would rather leave the detailed geophysical studies to the experts.

Old claim corners in Kantishna
with Denali in background

Chapter 20

Claim Staking

"Hey, you can stop what you are doing, the claimant is here," shouted Nathan. My wife's cousins and I were just finishing up putting in the third corner of a mining claim when I heard Nathan's shout. I bushwacked downhill to a clearing on a creek and introduced myself to the claim owner. He had a stutter which got worse when our staking crew started emerging from the bushes armed with shotguns and pistols. Earlier the same week, Nathan and I had researched an area where we thought would be a good place to stake some claims. We thought we were lucky to find an unstaked gold-bearing creek near Glenallen. The information we received from the BLM office showed the nearest claims located a mile south of the creek we wanted to stake. Nathan and I thought we should go out and stake this creek, so we would have a good place to recreation mine on weekends.

Since my wife's cousins were visiting from Minnesota and were two young and healthy boys, I invited them to help us out. On a Saturday in September, Nathan, the cousins and I packed up our claim stakes, 100 foot long measuring tapes, and geologic field gear and headed for the creek. We walked into the creek and found it had undergone a little mining. We thought we had lucked out by having such a nice place to mine. We put in our first stake and I told one of the cousins to take an end of the tape and walk towards a tree uphill from the stake. Claim lines have to be straight and since there was a lot of vegetation, the cousin had to bust brush to get to the tree. I kept him online with my compass and would stop him when I couldn't see him anymore. Whenever he stopped, he would put in a flag. I would tell the measurement to his brother and we would walk up and join his brother at the end of the tape. This procedure was repeated until we had gone 1500 feet in a more or less straight line. We then pounded in our second stake. We turned 90 degrees to the left and repeated the procedure for 600 feet. It was after pounding in our third stake while swatting at bugs and sweating profusely when we heard Nathan's call to abandon our claim staking effort. While walking back to the clearing through the brush, all I could think of was claim staking in a steep and brushy area was a lot of hard work, and I was kind of glad the work was over for the day.

After the claimant finally calmed down and found out we weren't a bunch of gun-toting claim jumpers, he told us BLM had misplaced his claims on their maps. Whoops! We asked if he would mind if we did some recreational mining before heading back to Anchorage. He told us to go ahead. Our mining effort gave the cousins a little gold and a good story to take back with them to Minnesota. But the story does not end there. Upon returning to work, Nathan and I were informed we couldn't stake any kind of mining claims because we worked for the BOM. We thought we were exempt from the non-staking rule because we didn't regulate mining. We thought wrong.

State and Federal governments allow people to stake mining claims on open lands. Prior to 1872, staking rules were usually dependent on the rules agreed to by miners in a district. There was no consistency between districts, which led to a lot of litigation. The Federal government finally decided to codify claim staking practices by passing the Mining Law of 1872. In general, the law allows a person to stake up to 20 acres of land, with a typical claim being 600 feet wide and 1500 feet long. Claims can be less than 20 acres, but not more. A stake has to be posted on each corner and a discovery monument has to be erected. The claim paperwork has to be filed with BLM and the State. A claim gives a person or entity the exclusive right to mine the mineral wealth on their claim. On the face of it, the mining law seems pretty straight-forward but because of nearly 150 years of litigation, there are complications and nuances associated with the law. Therefore, I'm not going any deeper into the law than to state a person can stake a claim.

Almost every field geologist I know has been involved one time or another with claim staking. It is hard work most of the time. Claim lines have to be run in a straight line for a set distance. Therefore, people staking claims have to overcome any obstacles in their paths and be able to compensate for any elevation changes they encounter. Ideally, people who stake claims should be able to read a compass and be able to climb up and down hills like mountain goats.

When I was working for WGM, they had an innovative method of staking claims in tundra areas. They would take the back doors off of a helicopter and fill the back of the helicopter with claim stakes. The stakes were 4-inch by 4-inch pieces of wood attached to a piece of rebar, which is just a metal rod. They would put one geologist in the back of the helicopter and another in the front. The helicopter would fly at a certain speed, in a straight line. The geologist in the front would have a map and a watch. Every minute the geologist in the front seat would tell the geologist in the back seat to throw out a claim stake. After so many corners were staked, the helicopter would turn, fly over to the start of a new line and fly in the opposite direction, repeating the process. After the job was done, the helicopter would fly geologists to the areas where the posts were thrown out in order to set the posts in the ground. Most of the geologists hoped the posts had set themselves by landing rebar first into the ground, which made their job a whole lot easier. A whole lot of claims could be staked in a short amount of time using this method.

I got involved with claim staking after I left the employ of the Federal government. To be honest, it's an activity more suited to young people who can scurry up and down mountains all day and not old people with bad knees.

BLM employees sampling bench gravels in Chicken

Chapter 21

Regulatory Work

One foggy and gloomy morning in 1978, Martin Maricle and I were driving a BLM truck down a muddy road near the Canadian border when we were flagged down by 2 men with machine guns. The road accessed Woods Creek, a tributary of the Fortymile River. Our job was to inform miners they had to reregister their claims with BLM by 1980. Martin was the "Chicken Man" who manned the Chicken Fire Guard station for BLM. He had volunteered to accompany me in the field and I was glad he was with me on this misty morning. Martin pulled the pickup over and the machine-gun toting men asked us what we were doing. In the back of my mind was the fact a miner had been ambushed and murdered in the Fortymile River area a couple of months before. We told them we were just visiting miners to inform them of the changes to the mining regulations. We then asked what they were doing and they said that they were rabbit hunting. Martin and I both thought it sounded kind of strange to go rabbit hunting with machine guns, but it was the Alaskan bush. After exchanging more pleasantries they said they were associated with the miner we were going to visit and if

I gave them the information they would let him know about the upcoming changes. I decided that the prudent thing to do was to honor their request and handed them an information sheet. After one of the men tucked the sheet into his coat he asked for a ride to the Taylor Highway. Since it seemed like the neighborly thing to do, Martin made a U-turn and told them to hop in the back of the truck. We dropped them off at the highway and went to find more of the area's miners.

Informing miners about changes in rules and regulations was some of the regulatory work I did in Alaska when I worked for BLM. Other regulatory work for BLM consisted of inspecting mining operations and conducting mineral examinations. While I worked for BLM, there weren't any strong regulations regarding reclaiming mine sites. If I saw a violation of mine safety or clean water regulations all I could do was report the miner to the appropriate authorities. In the 1980s, BLM beefed up their mining regulations.

I tried to visit all of the miners in the Fortymile River area. This was accomplished using 4-wheel drive vehicles and rafts. Since many miners were located along the Fortymile River, the easiest and cheapest way to visit them was using a BLM raft. I could easily persuade either firefighters or other resource people to accompany me on these information trips. We could get food at the BLM firebase, so costs were minimal. We would drive out to the South Fork Bridge and launch the rafts. We would then float down the river, talking to miners along the way, then continue to the Yukon River. We would float down the Yukon River and end the trip in Eagle where BLM had a fire station. Someone at the fire station would drive us back to our vehicles. It took 3 or 4 days to float the more than 120 miles.

While working for BLM, I evaluated 3 properties. I was involved with 2 mineral investigations in the Chicken area and assisted on a third in northwestern Alaska. The settlement of Chicken is well known in all of the literature about Alaska. It is famous because authors like to point out it got its name because the early miners in the area wanted to call their settlement Ptarmigan, after the game bird inhabiting the area. But they called it Chicken because no one could spell Ptarmigan.

The settlement consisted of a little store and bar with a couple of cabins scattered throughout the hills. I was fortunate to have worked in the area when some of the original miners and settlers were still alive. I was involved in examining a piece of land Ann Purdy's daughter wanted for a native allotment. Ann Purdy, who wrote the book "Tisha", came into the country as a school teacher, adopted some native children and never left. Her grown daughter had applied for a native allotment. Alaskan natives could apply for an allotment, but the land had to have no mineral potential. I was able to spend some time with Ann and feel privileged I got to listen to her stories about the good old days. Unfortunately, her daughter applied for a bench above Chicken Creek. Chicken Creek had been dredged, but the dredge could not mine the bench gravels because they were located high above the creek. To evaluate the bench gravel, BLM rented a backhoe for part of the work. The BLM District Geologist, Martin Maricle, another seasonal firefighter and I trenched and sampled the bench to determine its mineral potential. It wasn't too surprising gold was found in the gravels on the bench and therefore it had mineral potential. Instead of getting her allotment I heard Ann Purdy's daughter and her husband mined the ground when the gold price spiked in the early 1980s.

The second evaluation was a patent examination of a lode gold property in the Chicken uplands. At the time, a miner could apply for a patent to their mining claim. To patent a claim, a person had to first have the claim(s) surveyed by a certified surveyor, then file the proper paperwork. Once all of the i's were dotted and t's crossed, the BLM mineral examiner would sample the claim(s) and determine whether or not it was economic to mine. There has been a moratorium on patenting since the mid-1980s. The claim I investigated had little quartz veins or veinlets, which were less than 1/4 inch wide. One veinlet I saw contained visible gold. These veinlets were probably the source of the gold in Chicken Creek. Even though I saw gold in one 1/4 inch wide veinlet, I couldn't find any other gold on his claim. I left the BLM before the examination was finalized.

Besides meeting with Ann Purdy, I was fortunate to spend time with some of the "old timers" in the area. Billy Meldrum was one of the old timers I met in the Chicken area. He was in his 80s and still living full-time in a one-room cabin in Chicken Creek. He would hire a bulldozer for a day to strip topsoil from his claim, then in the summer after the ground thawed, he would mine the gravel with borrowed equipment. One weekend I took my father-in-law, who was visiting from Chicago, to meet Billy. During our visit, a friend of Billy's, who was also in his 80s, dropped by. The friend told a story of prospecting in the Dawson area. When we left Billy's cabin, my father-in-law said he liked hearing about the old-time adventures. I said, "Old time, he was telling us a story of what he did last week." Talk about tough guys.

Billy's friend was still working for a miner who controlled Lost Chicken Creek. The friend had owned the creek but sold it to a California rancher, who liked to mine in the summer. The rancher was mining the creek benches using a hydraulic giant. He had a pond constructed upstream and piped water down to a nozzle. Water would shoot from the nozzle onto the bank, like a fire hose. The water would wear away the overburden or frozen muck. When the overburden was gone, the gravel would be mined. The rancher invited me into his cabin and showed me some of the gold he had recovered that summer. He had mayonnaise jars full of a 1,000 ounces of gold. His mining also uncovered bison skulls, which he said he took back to California to line his driveway.

In 1979, I assisted geologists from the BLM Fairbanks office on their patent examination of the Arctic deposit. It is a large copper deposit in the Kobuk River drainage of Northwestern Alaska. The company which owned the deposit at the time had its own airstrip. We flew into the airstrip and rode in company vehicles around the property. The roads the company built around the property were paved with copper-bearing rocks. We didn't have to look too far to collect some nice examples of the rock they planned to mine. We spent a couple of days making sure the property had been surveyed properly, examined the discovery outcrop and oversaw some drilling taking place by the company for the patent examination. BLM required the company to drill holes into the deposit at certain spots to ensure the deposit actually existed at depth. The company met all of the criteria for patent, which the government eventually granted.

Fortunately, in 1993, I was able to return to the Fortymile River. I hadn't been back since I left BLM in 1980. I was investigating abandoned mines in the Fortymile Wild and Scenic River corridor for the BLM. The BOM had initiated an abandoned mine program in the early 1990s and this was one of the first studies undertaken in

Alaska for BLM. Since I was familiar with mining along the Fortymile River, I led the investigation. I convinced Nathan Rathbun and a BOM, Juneau office employee, to help with the investigation. We packed up a raft, canoe and 4-wheelers and headed for the Fortymile. I noticed how much had changed in the years since I had last been in the area. The guard station at Chicken was no longer a continuously manned fire station. It was run by the BLM recreation staff. The staff had stripped the station of all of the niceties present when I was working in the area, like electricity. There was an outdoor shower that was a welcome amenity. Also, all of the old miners I had visited in the late 1970s had died.

Our work not only involved investigating abandoned mines but also included looking for mercury along the river. Mercury is and was used by gold miners to collect free gold from sluice boxes or gold mills. As everyone knows, mercury is slippery stuff and has a tendency to get away from its confinement. Since the Fortymile River and its tributaries have been placer mined since the early 1900s, there was a good possibility free mercury was present in the gravels. 1993 was a good year to look for mercury because the area was in a drought and we could cross the river with 4-wheelers. This meant we could easily sample any likely spots on both sides of the river. Our panning efforts found mercury and a big cause of the contamination. The source of the mercury was traced to an abandoned bucket-line dredge on the Mosquito Fork of the Fortymile River, just downstream from Chicken Creek. A dredge is a self-contained mining machine that floats on a pond of water. It has a continuous line of buckets. The buckets can be as big as a bathtub and are attached on a boom, similar to teeth on a chainsaw. The boom is lowered into the water and the buckets scoop up the material on the side of the pond. The material then is deposited onto a rotating drum with a screen that separates the oversize material from the undersized. The oversized material goes onto a conveyor belt and it is deposited behind the dredge. The undersized material is washed into sluices where the heavy minerals are collected. Sometimes these sluices contained mercury or mercury was used after the heavy minerals were collected from the sluices. I thought finding one of the sources of mercury in the river was a significant discovery, which I reported to BLM. I don't know if they did anything with this information.

When I was a cheechako working in the Fortymile Resource area with a bunch of other 20-something cheechakos, the consensus was if the land was disturbed, it would never recover without the help of man. During our trip down the Fortymile Nathan and I stopped at numerous mine sites that had been stripped of all vegetation in the late-1970s. We found them choked with alders to such an extent it was difficult to survey the disturbed area. The natural reclamation that occurred in these disturbed areas showed me there are new things to be learned every day and the land healed quickly in Alaska.

Alaska has a term for people who live out in the wilderness, "End of Roaders". It seemed like the farther one got from the road system, the squirrelier people got. I don't think squirrelier is a word, but if the shoe fits... The miners, most of whom were using suction dredges, were OK if they were located on a road system, but there was a direct correlation between distance from roads and weird behavior. One of the first miners we encountered after leaving the last bridge on the Fortymile at O'Brien Creek started yelling at us and accusing us of spying on him because he had heard it via the Alaska "telephone" system.

We didn't try to reason with him because we were doing an abandoned mine study and his mine was definitely not abandoned. The last miner we didn't meet lived on Smith Creek on the Canadian border. He had a bear skull nailed to his front door. That Spring he had burned down one of his cabins when he thought summer had finally arrived. People said he was kind of surprised when it got cold again and he was without one of his cabins.

Fieldwork ended when we crossed into Canada. Nathan and I were looking forward to an easy float down the Yukon River. The Yukon flows at about 7 miles per hour, so I was used to just sitting back and relaxing, like Huck Finn on the Mississippi. This wasn't the case on this trip because the wind was blowing so hard up the river Nathan and I had to row in order to move the raft downstream. After fighting the wind for hours and getting nowhere, we finally gave up, went to a gravel bar, set up camp and went to bed. We had told people we would be in Eagle that day, but since there was nothing we could do and there was no way to communicate with the outside world we just made the best of a bad situation. In the morning the wind had died down and the float to Eagle was uneventful. Our only concern was making sure we were in the proper position to land at Eagle because the Yukon River at Eagle is about 1/2 mile wide and flowing fast. If we missed the town, the next stop was Circle which is hundreds of miles downstream. Luckily for us, we pulled into Eagle and were met with the BOM truck. We were a day late, which didn't seem to have concerned anyone.

In 1901, the town of Eagle was incorporated as the first city in interior Alaska. The city had formed around Fort Egbert, which was established to help the gold miners in the area in 1900. In 1903, a telegraph line between Eagle and Valdez was completed. In 1905, Roald Amundsen telegraphed the news to the world he had successfully sailed through the Northwest Passage. I only mention the telegraph line because when I was working as the Forest Geologist for the U. S. Forest Service in Sparks, NV, a woman was hired to temporarily help the Forest archeologist process mining permits. She was from Alaska working towards her archeology Master's degree at the University of Nevada, Reno. Her Master's thesis was on the Eagle telegraph line. After she was hired I found out that her father is/was a famous entertainer in Anchorage. While I lived in Anchorage I seen his shows many times. Small world. Presently, the town of Eagle has a permanent population of fewer than 100 people. The road to Eagle is only open in the summer.

Mark Meyer holding a nugget from a placer mine on Nolan Creek, Brooks Range

Chapter 22

Sample Analysis

Christmas in September, October and November. That's what I thought when the sample results started arriving at the office. I always loved opening the envelopes and going over the columns of analytical results to see if we had found anything good. After the fieldwork is completed and samples are sent to a laboratory, the laboratory processes and analyzes the samples to determine if there is anything good (i.e., economic) and how much good stuff is in them. I have been fooled many times thinking a rock has or has not a lot of good stuff in it just by looking at it. Therefore, I always send samples for some kind of analysis. Sample analysis can be done a number of different ways, but it's mostly done using chemical analyses. Chemical analyses started 4 millennia ago with people who practiced alchemy. Alchemy was and is a pseudo-science that was and still is practiced in various parts of the world. It involves chemistry and philosophy. The most common experiments by alchemists were those that tried to change base metals into gold. Even though these kinds of experiments failed to achieve their objective, the experimentation with various chemicals and methodologies eventually led to modern-day chemical analysis and refining of metals.

Today there are still what I would call alchemists and then there are legitimate laboratories. A person has to be careful not to send their samples to "alchemists" because once a chemical analysis has been done, it is hard to dispute the results. I had that problem when I started working for the BOM. The BOM had a multi-year contract with a low bidder lab to do all of the BOM analyses. Everyone knows sometimes low bidder and quality do not go hand-in-hand. And this was one of those cases. Luckily, some of the sample results were double-checked by the BOM Research Center labs and found to be so erroneous the contracted lab was only used for one year. The problem was I might have received bad results from this lab, but didn't know which results were bad. From that year on, I only used reputable labs, like Bondar-Clegg or Chemex. There have been so many changes with labs over the years I don't even know if these labs still exist, but at the time, they were the gold standard, so to speak.

I did get involved with "alchemists" who gave alchemy a bad name and who preyed upon unwary Alaskans. There was a guy in Anchorage who put together a one-page prospectus on his Kenai Peninsula claims. He would peddle his "investment" standing on the street corners of Anchorage. He was selling a square foot of his claims for $100. The claimant's prospectus stated a mineral deposit was under a glacier and he was going to dig down through the glacier and mine the rock. The prospectus also stated the mineral deposit contained a lot of platinum, gold and silver. The claimant had put together this prospectus based on the sample results he had received from his "partners". Besides giving analytical results, the partners convinced the claimant they could process

194

the ore from the mineral deposit and recover the precious metals. Their processing equipment consisted of a black box. They would put one of the claimant's rock samples in one end of the box and platinum, gold and silver bars would come out the other end on a conveyor belt. What exactly the machine did, was unknown to anyone but the "partners". People would come in the BOM office and ask about this "investment". Unfortunately, I couldn't tell them my honest opinion, which was it was a scam. All I could say because ethics prohibited me from giving my real opinion, was mining ventures were risky. I would then ask them if they could lose all of their money. I was stunned when people said they guessed they could lose all of their money.

I got more involved with this "investment" because the BOM was conducting a platinum study at the time and it would have been really great to have a major platinum deposit about 50 miles southeast of Anchorage. So, Bob Hoekzema and I went out with the claimant and had him point out the "ore" bearing rocks. He pointed to all the rocks laying in the creek below the hanging glacier. The rocks looked like barren sandstone and slate to Bob and me, but we took the rocks and brought them back to Anchorage. The claimant also pointed out the shiny minerals in the sandstone and said they were flakes of platinum. They looked like mica to me. I sent the samples off to a BOM lab to be analyzed. The lab concluded these were samples of barren rock. About the time the analytical results came back, the mining venture imploded. The partners with the machine took the claimant's money and their machine and headed south, leaving the claimant to deal with a lot of angry investors. Amazingly, the claimant had collected over $1 million, which was all gone when the venture came apart. The investors sued the claimant, but I heard he argued successfully in court he was stupid.

Similar scenarios play out all of the time where chemical analyses are involved. It's the old saying, "If it looks too good to be true, it probably is too good to be true." Most rocks have very low gold values, with most gold mines mining rock with less than $3 per ton of gold. If a lab reports multiple ounces of gold per ton in a sample, especially if there isn't any visible gold in the sample, then the lab results should be questioned. A lot of shady labs report multiple ounces of gold and platinum from the same rock. I have been working in the field for 40+ years and it is really really really hard to find any platinum except at established platinum mines. And rarely is there much gold associated with platinum. So, any lab reports with platinum should also be questioned. When I was working for BLM in 2002, BLM sent samples with known results to laboratories around the country. They received very mixed results. The results obtained by BLM research reinforced my opinion about only sending samples to labs I know are reputable.

Legitimate laboratories use only tried and true methods. They first take a sample, crush and grind it into a powder. A portion of the dissolved powder is put into a machine like a mass spectrometer or atomic absorption machine. The results obtained with a mass spectrometer are very comprehensive but are not as precise as those obtained using an atomic absorption machine. Because the results from a mass spectrometer are so comprehensive for many elements, it is a good and relatively cheap method to use when analyzing a stream sediment sample or just seeing what elements might be present in a rock sample. I call it the shotgun approach.

The results from an atomic absorption machine are

more precise than those from a mass spectrometer, but the labs charge for each element tested. If a person wants results using an atomic absorption machine, then the lab needs to be told which elements should be analyzed. A shotgun approach using atomic absorption can be very expensive. I would ask for atomic absorption analyses if I sent in samples where I could see copper, lead or zinc minerals. Atomic absorption will also give good results for gold and silver, but fire assay gives the best results for gold and silver. There are probably other machines and techniques labs are using, so I always talk to the lab prior to submitting my samples to see what methodology will give me the best results, not the highest, but the most accurate.

Sending rock samples to labs and waiting, sometimes months for results is the only way to determine if rocks have any good stuff in them. I didn't like waiting too much, so that is why I liked taking placer samples. Placer samples give immediate gratification. A person can pan out a sample of gravel, dirt or crushed rock and see if there is any gold, platinum or tin in any of the material. After a sample is concentrated by panning, the concentrated material (about half a sandwich bag of material) is put in a plastic bag and taken for further processing. Further processing consists of drying the sample, using a magnet to remove any magnetic material like black sands (magnetite), then using tweezers to pick out gold or platinum from the sample. The gold and platinum is then weighed on a precise scale to determine the amount of good stuff in a gravel deposit. The rest of the concentrate is then sent to a laboratory for analysis to determine what other elements might be present in the deposit or find any gold or platinum the tweezers might have missed. The recovered gold and platinum was then sent to a laboratory where it was analyzed for its gold and platinum content.

Raw gold is never pure and gold that comes out of a rock is the least pure. Gold eroded from a rock and washed down a stream usually becomes purer. Researchers have tried to map the gold content of gold particles in streams to try to determine how close they are to the source. Purities can have a big range. If my feet were held to the fire I would say the purity of hard rock gold is less than 90%, while the purity of placer gold is usually in the mid-80 to mid-90% range. It is important to know how much gold is actually in gold because a refinery will only pay for gold. So, if there are 100 ounces of gold at 90% purity, the refiner will only pay for 90 ounces.

Platinum is also not pure. A grain of platinum might contain various amounts of iron and sometimes valuable platinoids, like rhodium, osmium, iridium and palladium. After weighing the platinum grains, I would send them to a lab to be sliced and evaluated using a scanning electron microscope. A scanning electron microscope sends a focused beam of electrons onto areas of the grains and will give the chemical composition of those grains. It is an expensive and time-consuming procedure only used for special circumstances, like trying to figure out the composition of platinum grains.

In 1979, BLM sent me to a placer gold school. The school taught me to use mercury to recover placer gold. Mercury has been used to recover gold for centuries, if not millennia. The BLM placer expert instructed us to put a dime-sized bead of mercury in our hand and add it to a gold pan containing the concentrate. He told us we didn't have to worry about mercury poisoning because the oils in our hands would keep the mercury from being absorbed through the skin. He looked old and not crazy, so we all believed him. I've since questioned my san-ity for believing him. Gold dissolves into the mercury and forms an amalgam. The amalgam is then removed from the pan, mixed with nitric acid and the mixture heated. The mercury is burned off, leaving gold behind. We were warned to always stand upwind when burning off the mercury or better yet, use a fume hood. I used mercury a couple of times when I worked for BLM and the BOM. The last time I used it was in 1983 during the Kantishna study. In Kantishna we had a laboratory tent on site with a hood and exhaust fan so the mercury could be burned safely.

I stopped using the mercury method during the Kantishna study. I never liked this method and some of the placer gold in the Kantishna area was covered by a black coating (manganese). Because of the coating, using mercury to recover the gold particles was ineffective. Therefore, to make sure all of the gold was recovered from the concentrates, the concentrates had to be examined under a 10 power microscope and the manganese-coated gold particles separated by hand. Since all of the concentrates had to be examined using a microscope, it just seemed easier and healthier to skip the mercury recovery step.

Separating gold, platinum or tin using gravity methods is time-consuming. People are always looking for easier, faster methods. Mike Balen and I were going to conduct a placer investigation of the White Mountains near Fairbanks, where there are known tin occurrences. We were wondering how we were going to effectively separate tin minerals from the gravels. We knew the big tin mining operations around the world used jigs. A jig is a box with heavy balls on the bottom. Screened material and water are put into the box and the box is vibrated. Heavy minerals will sink to the bottom and pass through the balls. Light material will not pass through the balls on the bottom. Since jigs seemed to be very effective in separating light from heavy minerals, Mike and I wondered if we should buy one for our study. We contacted a guy in Anchorage who was selling small jigs and took a bag of tin minerals to test out his machine. We put the tin minerals into the machine and out the bottom came the tin minerals and gold. It seemed like this machine was kind of like the above-mentioned scammers' machine, but in contrast with the aforementioned "partners", the jig seller was an honest man. He told us he had let a friend of his use the machine to clean up his gold concentrates. He guessed some of the gold got stuck in the machine. He then told us about a construction worker who was involved with a building project in Anchorage. The worker had brought a bucket of gravel from his construction site, wondering if there was any gold in the Anchorage gravels. He put the gravels in the machine right before we ran our tin samples and was really happy when gold came out the bottom of the machine. He thought he had found a great gold deposit in the middle of town. Unfortunately for the worker and fortunately for us, by testing this machine we found it is really hard to prevent cross-contamination; therefore, we didn't buy a jig.

I ran into the same cross-contamination problem when Bob Hoekzema and I did some placer sampling in the Brooks Range. It was the first year of the Colville Mining District study and we borrowed a mini-sluice box from a BLM geologist. When we ran our first sample we were surprised to find gold. The only problem was gold was only found in the first sample and not in any of the other samples taken in the area. We concluded after talking with BLM, the sluice box had residual gold in its carpet from another BLM study. This emphasized the point that cross-contamination can lead to a lot of wasted time by geologists. Erroneous results also could have long-term consequences. One thing legitimate labs do is clean their equipment after every sample.

Abandoned Locomotive near Katalla

Chapter 23

Strange and Wonderful Discoveries

On a beautiful sunny day in July of 1982, Nathan and I were flying out to examine the Bering River coal field. While flying over a field of grass, we spotted something that shouldn't be in the middle of a field of grass, a locomotive. I exclaimed, "What the heck, over." The helicopter pilot asked me if I wanted to go examine this incongruous item and I said, "You betcha, set 'er down." This was one of the first really unusual items I discovered in the wilderness of Alaska, but it wouldn't be the last. By this time, I was used to finding "stuff" at mine sites. Mine sites usually have the typical unusual items miners would drag to their mine sites, but didn't want to drag back when they abandoned their hole in the ground. I have come across pickup trucks on the sides of mountains, dynamite, drills, drill rods, compressors, generators, and even placer dredges, but this was the first train locomotive I ever encountered.

I guess I shouldn't have been surprised by this discovery since it was located near Katalla. Katalla is south-

203

east of Cordova at the edge of Prince William Sound. At the turn of the 20th century, it seemed Katalla was going to be the locus of all kinds of economic activity. A massive copper deposit had just been found at the headwaters of the Copper River near Kennecott. Prospectors also were finding copper deposits all over Prince William Sound. A huge coalfield was discovered in the Bering River near Katalla. And, the first oil field in Alaska was found at Katalla. Therefore, the forces of industry thought Katalla would be a perfect place to build a copper smelter powered by either coal or oil. All that was needed to make everything possible was to build a railroad from the copper deposit at Kennecott down the Copper River to a man-made harbor at Katalla. The railroad would also haul coal for the smelter. Oil would be an added bonus. Copper would be smelted on-site and copper ingots and excess coal and oil could be shipped to world markets. Katalla would be the Pennsylvania of the west except instead of producing steel it would be producing copper.

Competing entities had all kinds of development plans for Katalla. However, their plans started coming apart in 1906 when President Theodore Roosevelt withdrew all Federal coal from mineral entry. Prior to this withdrawal, people could stake mining claims on Federal coal. Subsequently, in 1920, a law was passed that commodities like coal and oil had to be leased instead of staked, but that's another story. To add insult to injury, storms in 1907 destroyed an 1800 foot long jetty that had been constructed to form the harbor. The oil field at Katalla also turned out to be nothing more than a curiosity after a number of oil wells were completed and didn't produce much oil. Therefore, in two short years, it looked like the plan to turn Katalla into Pittsburgh was probably not going to happen in the near future. Therefore, to exploit the large copper deposit up the Copper River, in the shortest possible time, the copper company abandoned their plans for Katalla and moved the terminus of their railroad to Cordova and built their smelter in Tacoma, Washington.

Some entities still had dreams for Katalla once the Federal government solved the coal question. One entity acquired the coal rights to the Bering River field and proceeded to build a railroad to tidewater. Once the railroad was completed, it started exporting coal to the continental U.S. However, the company quickly found Alaskan coal couldn't economically compete against the coal fields in Wyoming, Colorado and Utah. Therefore, they left and abandoned their equipment to the Alaskan wilderness.

The Katalla area had more surprises in store for me. After a couple of minutes for photo ops around the locomotive, it was off to the coal fields. However, Mark Meyer, who was in the area with our other helicopter called and said, "Hey, ya gotta fly over the nose of the Bering Glacier." So, we deviated from our flight path a short distance and saw a phenomenon of nature. There was a geyser of water shooting straight up from the ice. It seemed a blockage under the glacier was diverting the river under the glacier straight up through a crevasse. It was amazing to see multi-thousands of gallons of water shooting 30 feet into the air.

Another surprise occurred when my boss, Don Blasko came out to Cordova for a visit. He was a petroleum engineer and was very familiar with the Katalla oil field. He took my crew out to sample the oil seeps. We were all hard-rock geologists who never worked in the oil field. He showed us where the oil seeps were located and how to sample the gas bubbling up through the oily water. The surprise came when we were sampling the gas. Frogs would jump into and out of the water. It seemed to us the frogs had found a good way to protect themselves from predators by covering themselves with oil. The oil didn't seem to have any ill effects on the frogs because they were prolific.

The Copper River area had some other noteworthy aspects. One was the Million Dollar Bridge. It was built across the Copper River at the mouth of Miles Lake. It got its name because, in the early 1900s, it cost a million dollars to build. This is when a million dollars represented a lot of money. The bridge was built to handle the rail traffic between the Kennicott (also known as Kennecott) copper mine and Cordova. When the railroad was abandoned, the bridge was used for vehicle traffic. When I was working out of Cordova in the early 1980s, we could drive as far as the bridge, but wouldn't cross it because the decking on one part of the bridge had fallen into the river. Some handy person or persons had cobbled a couple of planks across the abyss. People would drive on these planks, but I wouldn't risk government equipment on the bridge. Plus I had a helicopter.

Another noteworthy attraction in the Copper River area is Kennicott or Kennecott. Kennicott is an old mine and mill site in the Wrangell-St. Elias National Park and Preserve. It consists of the company town of McCarthy, plus old mill buildings near the mine. The 2010 census indicated that there were 20 people living in McCarthy. When I was living in Alaska the mine and mill site was owned by a private company. Much of the property around Kennecott was acquired by the National Park Service in 1998. One weekend, Bob Hoekzema, Nathan Rathbun, Joe Kurtak and I, plus some other friends made a field trip to the Kennecott Mine. We loaded up our camping and field gear and drove to McCarthy. In the 1980s, in order to get to McCarthy, a person had to park on the bank of the Kennecott River and load their gear onto a hand tram. A hand tram was a platform with two seats hung from a cable stretched across the river. Two people would get on the tram and pull themselves and their gear across the river. Once across the river, we bummed a ride in the back of a local's pickup to the mill. We then proceeded to set up our tents on the roads surrounding the mill and started exploring the area.

The Kennecott mill is one of the most impressive abandoned structures in Alaska. It is multi-stories tall and painted red. It literally hangs on the hillside above the Kennecott glacier. The weather was perfect for our foray. We walked out onto the glacier and all through the mill. Most of the equipment was still in the buildings. The next day, we packed up our underground mine gear (i.e., lights and hard hats), plus our rock hammers and hiked the 4,000 vertical feet to the main mine opening. We had heard and read the mine tunnel went through the mountain. Our friend, who came with us, told us he had walked the entire length of the tunnel and would lead the way. Since we considered ourselves experts in exploring underground workings, we thought it was going to be a piece of cake. Upon entering the mine opening, we were awed by the copper minerals we found right off the bat and we thought this was going to be a great adventure. However, our little adventure was cut short way too soon when we encountered an area where the unstable rock had fallen and caused the tunnel to be blocked by mine timbers. When we pointed out the fallen timbers, our friend/guide told us this was only the first of many similar problem areas in the tunnel. He also said we could probably crawl through the timbers as long as we were very careful not to touch anything. Well, we had a confab right there. We decided we could probably make it through this problem area without dying or getting trapped. But since we knew there were many more areas of unstable ground in the next nearly mile of tunnel, we decided to not chance it. Therefore, by consensus, we decided a not too hasty retreat would be the best alternative and wouldn't result in anyone getting seriously injured. We were able to collect some nice mineral specimens and had a nice hike down the mountain. I think McCarthy and Kennecott have completely changed since I was there in the 1980s.

People who want to understand the early development of Katalla, Kennecott and Prince William Sound should read "The Copper Spike" by Lone E. Janson. Rex Beach's "The Iron Trail", is a fictionalized version of these events.

I have also found some strange things in the bush. While walking down a creek in the Brooks Range, 100 miles from the nearest village, I spotted something red lying in the creek. My field partner and I came up to this red thing and found it was a mylar balloon in the shape of a heart, which said Be My Valentine. The balloon was still partially filled. To think someone somewhere got a mylar balloon on February 14th and it got away from them, then it floated hundreds of miles to land in a creek and I found it, blew my mind.

Another mind-blowing discovery was an airplane sitting on the banks of a lake on the North Slope. While working on the north slope in the early 1990s, we heard about a plane that had problems getting off of a semi-frozen lake. The plane, a DC-3, had been ferrying in supplies to a government drill rig in the winter. It had successfully landed and taken off from a frozen lake numerous times. However, Spring arrived a little early and after unloading its cargo, the plane tried taking off over the mushy surface of the lake. Its landing gear fell through the ice and that is all she wrote, so to speak. A salvage crew later came to the lake, dragged the plane onto the shore and stripped the plane of its electronics. So, here it sits 30+ years later, on the edge of No-Luck Lake, like some big aluminum lawn ornament.

There were other strange discoveries in Alaska, which I really treasured, but they were built for the tourist trade. There is a 40-foot tall Igloo along the Parks Highway about halfway between Anchorage and Fairbanks. It was going to be a hotel but was never finished. There was the Bird Creek Saloon along Turnagain Arm, which was festooned with a big blue bird with an orange beak peeking out of the attic. The saloon was a log cabin that had settled into the tundra at an angle. A person had to hold onto their drink so it wouldn't slide off the bar. One of the bartender's favorite jokes was to have people blow their ptarmigan whistle. The whistle was a curved pipe with white powder in it. A person would blow into one end of the pipe and white powder would shoot out the other end into the person's face, which produced a good laugh from the all-knowing bar patrons. Over the years people had tacked all kinds of things, like money, panties, brassieres and business cards on the walls of the cabin and it became a fire trap. It eventually succumbed to an errant ember. Then there was the famous Bucky the Moose in Tok. Bucky was a stuffed moose located near the intersection of the Glenn and Alaska Highways. People would get their picture taken riding Bucky. It finally succumbed to the elements and excess love.

These were and are only a few of the unusual things a person can find, if they get out and about in the Alaskan bush. I was always a little awed when I encountered these odd discoveries.

The author and fellow USBM employee wearing headnets in order to sample an outcrop in the Brooks Range

Chapter 24

Bugs

"Thirty-five, thirty-six, thirty-seven, thirty-eight, thirty-nine, forty, top that," I said to Tom as we sat waiting for the helicopter along the banks of the Yukon River. He slapped his thigh and started counting the dead mosquitoes. We were having a contest on who could smash the most mosquitoes in a single slap. The record by our field crew so far this summer was 100, but since it was getting late, it didn't look like we were going to break the record before the helicopter arrived to take us back to camp. It did, however, look like Tom was going to have bragging rights this day. He had just smashed 50 when we heard the beating of the helicopter rotors and decided we had better gather our field packs and hold onto our head nets.

One of the biggest hazards to fieldwork in Alaska, besides the steep slopes, are bugs. Alaska has three kinds of nasty bugs. One is the ubiquitous mosquito. Nearly everyone who has written about Alaska writes about swarms of mosquitoes. The swarms have been linked to causing caribou to stampede. I'm not a biologist so my mosquito observations might not be scientifically correct, but I and many of my colleagues have concluded the first mosquitoes of the summer are big and slow, which makes them easy to swat. As the summer progresses, it seems the mosquitoes become smaller and faster, which makes swatting them more challenging. Mosquitoes also seem to be more prevalent in the interior than along the coast of Alaska. The Yukon River region is really thick with them, whereas in the coastal regions it rains more and is usually windy, so it might just seem like there are fewer mosquitoes.

The white sock is another bug that terrorizes most bush Alaskans. It is a small black fly with white feet, which makes it look like it is wearing socks. White socks usually emerge in mid- to late- summer. They like to fly into mouths, noses and eyes. They will bite any unprotected piece of flesh. They climb up gloves or socks and bite where the long sleeves of a shirt or socks end. Many geologists either tuck their pants into their rubber boots or wear gators over their hiking boots to prevent these bugs from climbing up their pant legs.

The last nasty bug in Alaska is the No-See-Um. Its scientific name is Ceratopogonidae. It is a very small bug, thus the name, with a nasty bite. When a No-See-Um bites me, I know it. The area around the bite will swell up to a baseball size lump and itch like crazy for days. They emerge about the same time as the white socks and therefore they deliver a one-two punch to an unsuspecting animal.

Luckily, I was born in an era where people developed ways of thwarting these menaces. I can't imagine working in Alaska prior to the use of mosquito or bug repellent. The old-timers coated themselves with grease, used smoke or used head nets to combat mosquitoes and other bugs. Now, modern people use sprays and liquids containing DEET. DEET is diethyltoluamide that is added to a liquid. It is very effective. Since while working in the bush I would only take a bath or shower about once or twice a week, I was pretty immune to bug bites until I washed the bug dope off. On some occasions, I used military bug dope. It had a much higher DEET content than commercially available brands. It came in clear bottles and was kind of oily. The interesting and I use the word loosely, aspect of the military bug dope was it would dissolve plastic and strip paint. So, I tried to keep my hands away from my plastic eyeglasses and plastic lenses. However, since I had to handle my pens and pencils, they would usually not be in pristine condition after using military bug repellent. It was very effective in keeping bugs away, but I didn't particularly like using it because I always wondered if it had such adverse effects on inorganic substances, what were the effects on organic substances like my skin and anything below my skin?

Everyone wants a less obnoxious repellent that is more "organic". I say, "Go for it, good luck." I'll stick with tried and true chemistry and lots of it. I'll spray my hat, head, arms, hands and clothes. Then I'll have a backup bottle in my field vest, just in case the spray can runs out of juice. I've seen some people get so frustrated with the bugs they'll grab their spray can press down on the nozzle and light the chemical stream with a lighter, making a flamethrower. The only good outcome from such a maneuver is a temporary sense the person is taking the battle to the bugs. However, the bugs always have the last laugh because the maneuver uses up the repellent and the supposed aggressor is now left defenseless against the ensuing onslaught.

My other defense against the wily bugs was a head net. There were two types. One has a solid nylon top with a metal ring around the top of it and netting on the sides. The metal ring is about a foot in diameter and

keeps the mesh away from the skin. There is a draw-
string at the bottom of the mesh to close off the bottom.
I could wear this head net as a hat if I didn't have a hat,
but the mosquitoes could drill through the nylon top into
my head. Therefore, because I found out the itchy way
that a head net doesn't make a good hat I carried multiple hats. The second type of head net, which I always
carried in my emergency kit, was just a mosquito net I
could pull over my head. When using one of these nets it
was advisable to have either a full-brim hat or at least a
baseball cap to keep the net away from the head. I never
knew when I would need either of these nets, so I always
tried to have them in my backpack. There were many
times waiting for a helicopter I would put on my head
net and rain gear so the bugs wouldn't drill through my
clothes. I would sit patiently and watch the dirty buggers
crawl all over me trying to find any chink in my armor. I
always had to make sure to take off the head net before
approaching the helicopter or I would see it torn off my
head and flung into the netherworld by the rotor wash.
Thus the need for a backup head net.

During my first summer in Alaska, I was winning
against the bug assaults by employing plenty of repellent and head nets until I camped at Lake Sithylemenkat.
Lake Sithylemenkat is a fairly circular lake, which is
approximately 2 miles in diameter. It has been postu-
lated that it is a meteor crater. Whatever it is, the lake in
the summer is ideal for swimming and/or bathing, even
though it is located just south of the Arctic Circle. Since
I badly needed a bath and hadn't gotten bitten for weeks,
I thought I must be immune to the bugs. I was also sick
and tired of constantly being covered with bug dope and
just wanted to be squeaky clean for once. So, I plunged
into the lake with my bar of soap and bottle of shampoo.

When I was finally clean, I climbed onto the beach and let the sun dry me off. What a euphoric feeling to be free of days of dirt and oceans of bug dope. However, the euphoria was short-lived because I woke up in the night itching, scratching and swollen from the white socks and no-see-um bites. This was a good lesson in realizing that the bugs always win unless a person is prepared. It was also the last time I didn't wear bug dope in Alaska. Like I said some places were worse for bugs than others, but I was always prepared for the worst and tried not to let my guard down.

Bear in Denali National Park

Chapter 25

Bears and Other Beasts

No book about Alaska would be complete without a bear discussion. Black, grizzly or brown, and polar bears are a constant concern in Alaska. People look at the Kodiak brown and polar bears and think Alaska is filled with huge bears. In reality, the huge brown bears are found on Kodiak, hence the name, and polar bears are along the coast of the North Slope. I never encountered either of those bears; however, my wife who worked in both places had some encounters but those are her stories. I spent my career in places where the bears were a lot smaller. The weight of a black bear in interior Alaska is less than 200 pounds, which is about the size of a large dog. The weight of an interior brown bear or grizzly has never been scientifically determined but is generally believed to be less than 500 pounds. Even though these bears are not huge by Kodiak or polar bear standards, they are still dangerous.

I came to Alaska wide-eyed and stupid. My first night in Anchorage started my Alaskan educational experience. I and the other seasonal WGM employees were sitting around solving the world's problems when the experienced hands started telling their bear stories. One guy thought of himself as a bear magnet. Wherever he camped, he always set up his tent next to a climbable tree because almost like clockwork, he would be woken up by a bear in camp and he'd spend the rest of the night in that tree. The bear stories continued throughout my first summer and culminated when a female geologist working for USGS, who wasn't carrying a gun, was mauled in eastern Alaska. Her injuries were so severe that both arms had to be amputated. I ended up running into her, with her prosthetic arms, throughout my career. Small world. By the end of the summer, I was really freaked out about bears and would borrow any available gun before venturing into the field. My concern was reinforced when on one day I ran into six grizzlies. It seemed like every place I either wanted to sample or wanted to drop other people off to sample, there was a bear. Sometimes I didn't see the bear until after taking the sample. One time, I was going to drop someone off and spotted a bear stalking a moose at the proposed landing area. For some odd reason, my co-worker refused to get out of the helicopter and I had to find another spot for her to work.

While I was living in Anchorage in the fall of 1977, the other seasonal geologists and I would go out to the Anchorage zoo and watch the bears. We would just stand on our side of the fence for countless minutes watching these animals do what they do. I guess watching the bears was so fascinating to us because it was one time we could watch bears safely without adrenaline pumping through our bodies and wondering whether or not an attack was imminent.

I was lucky with all my bear encounters. Most times it seemed the bears were just curious about an interloper in their backyard. An observation I made while working in the Brooks Range was the bears west of the Dalton Highway would just look up at the helicopter as it passed over them. The bears east of the Dalton Highway would run if they heard a helicopter. It is my opinion the bears west of the Dalton had hardly ever seen a helicopter because few wildlife studies had been conducted in that remote area, whereas the bears east of the Dalton Highway near the Arctic Wildlife Refuge had been poked, prodded and tagged by researchers and therefore associated helicopters with bad things happening to them.

Other beasts in Alaska a person might be concerned with are wolves, moose, caribou, wolverines and rodents. It seems wolves have been part of a person's psyche since Little Red Riding Hood and the three little pigs had their encounters. I know there are wolves in Alaska and have heard many stories about wolves, but never saw one, except in the zoo or as a pet. They are big animals and I wouldn't want to run into one unarmed. While living in Tok, my wife saw a mama wolf and her 9 pups crossing the Taylor Highway. The local Alaska Department of Fish and Game biologist was upset when she told him about the prodigious number of pups. For decades Alaskan biologists had tried to keep the wolf populations low because wolves compete with humans for the same meat sources. The thought of nine more wolves in the biologist's domain was disconcerting.

Alaskan moose are huge. The males have big antlers and all moose have razor-sharp hooves. They are an animal a person wants to avoid unless they are moose hunting. It seems people get injured or killed by run-ins with moose yearly, but those occurrences are usually in an urban setting. One time my wife and I were in Alaska visiting friends and counted 11 moose in a couple of weeks. During that trip, the moose we spotted were either in

Anchorage or in other towns. The moose hang out in the cities because people plant tasty vegetation in their yards and they are protected from hunters. Therefore, the likelihood of a human-moose encounter is higher in an Alaskan city than in the bush. When I encountered moose in the field, I tried to keep a lot of vegetation between the moose and me while making a wide detour.

Caribou are essentially wild reindeer. Biologists might disagree with my assessment, but I'm a geologist. They are smaller than a moose, but also have some defenses against natural predators, like antlers and hooves. It seems caribou are not very aggressive. The ones I've encountered just stared at me or walked away. But, like any animal bigger than me, I'm wary of encountering them in the bush.

An animal with a bad reputation is the Wolverine. It's kind of like North America's version of the Tasmanian devil, which is an animal walking around with a chip on its shoulder. Wolverine might have a bad reputation but are rarely spotted. I count myself lucky I saw one in my years in Alaska. I was floating down the Fortymile River and one swam across the river in front of my raft.

Rodents are more of a nuisance than a danger, especially in camps. It seems geologists like to camp in the same spots year after year. Because there is always a lot of food around camps, the rodent population seems to multiply accordingly. Therefore, after a number of years, the population gets so great they have to be dealt with or everything will be chewed up. In one camp, it got so bad everyone got their guns, then put food on the ground away from camp and waited for the rodent hoard to arrive. Once there was a feeding frenzy going on, everyone started blasting away. I don't know if it solved the problem, but it made everyone feel a little better fighting back against the voracious beasts.

I dealt with bugs by wearing bug dope. Whereas, I dealt with beasts by wearing a gun. When I found out I would be returning to Alaska in 1978, I went out and bought a Ruger .45 caliber revolver. I subsequently bought a shotgun. I carried these for personal protection. At the time, Alaska was very progressive when it came to gun-toting. I could walk the streets of any town and into any business with a gun and no one would blink. So, carrying a gun in Alaska became just a way of life. For work, I carried either a government supplied Ruger .44 pistol in a shoulder holster, a Smith and Wesson .44 pistol in a hip holster or a Remington 12 gauge shotgun. I only carried a shotgun in grizzly country. To this day, I don't know if any of those guns would have stopped a bear, wolf, moose, caribou or Wolverine and luckily never had to find out. I think I carried a gun because it was like a security blanket, knowing I could try to defend myself if something bad happened.

Pistols are easier to carry because I put them on just like my field vest and belt. I didn't have to worry about them getting wet if it rained or getting the barrel full of bushes or water. The shotgun was more problematic because it had to be slung over the shoulder with one hand holding the butt of the gun. I would put tape over the barrel to keep things out of it. There was a debate on what ammunition was better for the shotgun, either double 00 buckshot or slugs. The common logic was to load the shotgun with a slug cartridge, then alternate with a double 00 buck cartridge. The logic was that if the slug didn't kill bear the double 00 buckshot would blind it. When Sabot rounds became common, I just carried those. Sabot rounds supposedly could stop a charging bull, so they seemed like they would be good enough to stop any kind of wild beast. If I carried a shotgun, I couldn't carry something else, like maps, clipboard or rock hammer. So, every time I needed one of those items, I had to set everything down and get it out of my vest or pack. Therefore, I really didn't like carrying a shotgun unless I thought I needed extra stopping power.

One common problem I had was loading and unloading the gun when helicopter-supported. It was against the rules to carry a loaded firearm in a helicopter. So, in the field, I would wait until the last minute to unload the gun because I didn't want to be surprised by a bear right before pick-up. When I got out of the helicopter, I reloaded the gun before starting fieldwork. This was a real hassle, especially if I thought I would only be at a spot a short time. So, many times I wouldn't load the gun in those situations and inevitably I would run into a bear. One time I was let off on a rocky area next to a glacier. I thought, "Why load the gun, I'm on a glacier?" And, what do I see walking up the glacier a couple of 100 yards from me, but a black bear. So, I started retreating, yelling and frantically loading my gun. Another time, it was a hot day. Nathan and I were taking a placer sample. We had stripped down to our tee shirts and put our guns on our packs. While I was panning the concentrate, I looked across the creek and there was a bear staring at me through the alders about 10 feet away. I yelled at Nathan, "Bear!" and we both jumped to get our guns. Luckily the bear seemed only curious about these skinny hairless creatures in his backyard and assuaging his curiosity went back to his wandering.

Encountering wildlife in the field always caused a spike in adrenalin. I was always hypersensitive when walking down a trail because I knew no human trail crew had made the trail. Some of the trails along creeks and rivers were lined with grass 4 feet tall and contained the remnants of some animal's breakfast, lunch or dinner. I knew because the grass was so tall, I was probably going to surprise any beast using the trail. Even though I was worried about running into a beast, especially of the bear-variety, I never let the worry overwhelm my sense of getting the job done. I would just forget about the last encounter until the next encounter happened.

There was always dark humor in camp about bears. People would joke one should always work with a partner who runs slower than them. Or, people would joke if only one person had a gun and the partner didn't, the person with the gun didn't have to shoot the bear, all he or she had to do was shoot the other person in the leg and run. The second joke was just a joke and a not too funny one at that, but the first joke has more truth to it than people would like to admit.

Joe Kurtak admiring the view from a camp site in the Talkeetna Mountains

Chapter 26

Entertainment

"Hey, who tied a salmon to our bumper?" I shouted to Mark as I was attaching the tow rope to the rear of the BOM truck and he was securing his end to the other BOM vehicle. While driving back from the field on a dirt road outside of Cordova, the driver of the lead truck got too close to the edge of the road. The shoulder of the road gave way and sucked the right side of the truck into a ditch. We were able to yank the truck out of the ditch after untying the fish. Before we flung the fish into the tundra, we noted the fish was still in pretty good shape because July had been so cold and wet. Everyone in my crew denied the prank and it wasn't until September when we were back in the office Nathan asked if we had discovered the present he gave us before we left for the field in June. I think he was kind of disappointed when we mentioned we hadn't noticed his present until there was a reason to look under the bumper, which made his prank less entertaining than he had hoped.

Entertainment in the field was different than in camps. There was always some kind of entertainment in a field camp, even if it was just playing cards or sitting around trading "true" stories. But, while working in the field, geologists had to make up their own entertainment to pass the hours of slogging and banging on rocks. Entertainment in the field is not the same as in an office environment, where there are birthday and retirement parties, and "water cooler" discussions.

There were a few basic ways geologists entertained themselves. One was saying dumb things. There were some common geologist expressions usually said to seasonal employees when the newbie geologist would ask about a rock they were holding. The replies were: don't take it for granite; it's a gneiss rock; it's leaverite, which means you leave 'er right there; or it's a sex stone, which stands for another friggin' rock.

Another form of amusement was pulling pranks like the fish gag. The most common prank was messing with a field partner's pack. This usually entailed transferring rock samples from one's own pack to someone else's pack. This prank was always good for a laugh when the partner finally figured out why their pack was so heavy. The pranker hoped the prankee wouldn't figure out until they got back to camp. Sharing the prank with fellow campers made the prank much more amusing.

A field partner pulled a prank on me one day when we were looking for rocks that contained scheelite. Scheelite is a tungsten mineral that glows blue under a black light. My partner and I had a black tarp and a black light. We would crawl under the black plastic tarp, so it would be as dark as possible and shine the black light onto the rock samples. Since it was a blue sky, hot day, no one wanted to be covered by black plastic and lamp all of the promising samples we had collected. But after much hemming, hawing and flipping coins, I won the honor of being encased in black plastic while my partner handed me rock samples. The first couple of minutes weren't too bad until the sweat soaked with bug dope started streaming down my brow and into my eyes. And, to my chagrin, the rocks just kept coming and coming. Who knew that we had collected so many promising specimens. It wasn't until my partner couldn't stand it anymore and started laughing that I suspected I was a victim of his tomfoolery. He, of course, had been handing me the same rocks over and over and because it was so dark under the plastic I couldn't tell the difference.

Pranks were not very sophisticated nor very funny, but they did break up the day. I'm sure other geologists have pulled some very good pranks they will share if asked, but that was about the extent of my involvement with pranking. The reasons for quitting the pranking business were that I was afraid to pull pranks on the boss and when I was boss, I never wanted to pull pranks on my helpers. If I pulled a prank on the boss, then the boss could always retaliate by sending me on a "death" march. A "death" march didn't mean the person would die. It was just a day that included the hardest nastiest work the boss could devise. And, if the boss pranked their assistants, then the assistants might get their noses bent out of shape and not work as hard. Therefore, pranks were mostly between seasonal employees or between permanent employees. When there were pranks between permanent employees, everyone realized there would be a chance for a retaliatory prank sometime in the future.

Another form of entertainment was rock rolling. It is a tradition when a geologist either reaches the top of a hill or is working high on a slope, they find the biggest rock they can move, then they push it and let gravity take over. It's hard to explain, but there is a certain joy or satisfaction in seeing an object bouncing down a slope, picking up speed then crashing at the bottom. It's especially satisfying if there is a body of water so the object makes a big splash. Probably the most interesting rock rolling experience I was involved in was when Nathan Rathbun and I were working on a steep slope on the Kenai Peninsula. We got a big rock rolling and it bounced and bounded down the slope and crashed into an alder patch. Well, out of the alder patch comes this big black bear running for its life. I imagine the bear thought his nice world where nothing fell on him was coming apart and the better part of valor was to make a run for it. Nathan and I thought, cool. Then we thought, boy it's lucky we didn't walk through that alder patch on the way to examine a prospect. We were also happy the bear decided to head downhill instead of uphill.

A serious form of entertainment was target practice with a pistol. Since I only handled a .44 caliber pistol in summer, with target practice usually limited to about a week during safety training, I felt justified in using downtime in the field to sharpen my skills. It wasn't a daily occurrence, but every now and then after lunch, I would unlimber my firearm and plink at an object. Plinking with a .44 is not really an accurate term. The magnum ammunition used in a .44 makes a huge bang and the recoil does a number on the hand and forearm with each shot. Not too many people shoot these pistols with this ammunition for fun, so I limited my practice to a couple of shots at usually a piece of rotten fruit I had for lunch. Because there is such violence associated with shooting these weapons, practice for prolonged periods usually didn't increase accuracy, it just made me start flinching because I was anticipating what was going to happen when the trigger was pulled. I just took some shots every now and then because I wanted to keep my skills up and to make sure I wouldn't flinch if I had to shoot at something, like a bear.

It was always important, even in the middle of nowhere, to be aware of where the bullet might go when the trigger was pulled. This was never more apparent than the day I was working along the Copper River near Cordova. I had finished work and was waiting for the helicopter. I thought I'd do some target practice to pass the time. I took aim at some bushes, then thought better of it because I really didn't know what was on the other side of the bushes. About that time, some hikers came out of the bushes I was targeting. What the hikers were doing walking in some bushes around the Copper River, with no vehicles in sight, I never knew, but that incident reinforced the practice of knowing exactly where the bullet might go and that Alaska isn't as empty as people think.

Dumb expressions, pranks, rock rolling, rock throwing and shooting guns seem pretty juvenile. I can't disagree with that assessment, but it was a way of breaking up a day.

Helicopter pilot Ralph Yetka conducting a maneuver at a prospect on the Kenai Peninsula

Chapter 27

Safety Awareness, Training, and Equipment

"OK, Nathan. Bob said all we have to do is walk across this snow slope and climb up to the building where the mine is located," I said when the helicopter let us off on a flat spot high up on a mountain slope. This was going to be a piece of cake. We only had to walk 1500 feet over a snow slope and map and sample a mine. This mine, when active, reported gold stringers hanging from the ceiling after each blast. I thought I was lucky to get this assignment because I was the "new" kid on the BOM crew. I had only worked for the BOM little more than a month and felt like I was still proving myself. "Alright Nathan, let's go do this," I said as I stepped on the snow slope and slipped on a sheet of ice. I took a step back and thought about the situation. It was early morning and we were on a north-facing slope. The sun wouldn't hit the slope until the afternoon. The sun might soften up the surface of the snow in a couple of hours, but maybe not. What were we to do? I looked around and did a mental inventory. We had our field gear and a shovel. We were wearing rubber boots, which didn't provide very good traction on ice. In reality, there was really only one option for the "new" kid. I grabbed the shovel and start cutting footsteps in the ice. After cutting 20 or 30 steps, my legs started shaking as I looked down the slope of the ice. The ice slope was at a 45-degree angle. I thought while looking at it, if I slipped, the 1,000-foot elevation drop to the rocks below wouldn't be a fun ride. My nerves were still raw from a near-death experience I had on a snow slope in Colorado the year before, so I decided it was Nathan's turn to have a go with the shovel. After another look down the slope, I retreated to the flat spot where Nathan was waiting. I handed him the shovel. He walked out and cut steps until his nerves also gave out. He then came back to the relative safety of the flat spot and handed me the shovel. I didn't feel so bad about my weakness after someone else had the same anxiety I experienced on that slope. I took the shovel and resumed cutting steps where Nathan had left off until once again, I couldn't do it anymore. We proceeded in this manner for a couple of hours until we finally reached a ladder leading up to a building. The miners had attached a building to the vertical rock face and they accessed the building via the ladder. I climbed the ladder and went through a trap door in the floor of the building. To my surprise, I found dynamite and blasting caps strewn on the floor. Piece of cake, my eye. While staring at the dynamite, I saw out of the corner of my eye a truck sitting in the shack. I guess the miners or space aliens had hauled up a truck and installed it to provide power for the mine. Will wonders never cease? I told Nathan to come up and be careful. We both knew we could have called it off right then and there, but after spending all of the time it took to cut steps and feeling kind of wimpy, we weren't going to let a little bit of dynamite stop us from getting the job done. We tiptoed through the dynamite (sounds like a song) and out of the building to a path cut along the rock face. We followed the path over to the mine opening. To our disappointment, we found a big block of ice from floor to ceiling at the entrance to the mine. "Oh shucky darn," we proclaimed or some such expletive. We then retraced our steps and called the helicopter.

I relate this story as a teachable safety moment. Here were two people in their 20s who it seemed were not very safety aware. Why weren't we prepared with ice axes, ropes and crampons to tackle an ice slope? Why did we gingerly navigate a floor strewn with dynamite? I attribute it to culture. When I first started working as a geologist, in the late 1900s, the attitude was to do whatever it took to get the job done and hopefully not die doing it. Employers reinforced this attitude by giving an employee freedom to work with little or no oversight, as long as the freedom didn't negatively impact the bottom line. Therefore, my early jobs didn't require any type of safety training nor did I expect it. Younger people working now might view workplace safety as just an inherent part of their job, but it wasn't always so. In the 1970s Congress passed workplace safety laws like the Occupational Safety and Health Act of 1970 and the Federal Mine Safety and Health Act of 1977. It took decades after the passages of these acts before their impacts were felt in the workplace. Ironically, Federal agencies were exempt from these laws.

When I was in graduate school at CSU in the mid-1970s, the geology department had an X-ray machine used by students to identify minerals. With very little training, students could prepare their samples and use the machine. A couple of years after I left CSU the university started implementing safety rules. During a safety check of the x-ray machine, it was found the machine leaked and shot x-rays into the hallway. I wonder to this day how many x-rays I absorbed walking the hallway.

Also, in graduate school, I had free rein of the labs and its chemicals. Many of my fellow students, myself included, suffered the ill effects of acid burns and toxic fumes, while working in the labs. Also during my time in school, I spent months in the field by myself with no oversight. The lack of safety rules might seem horrendous and stupid in today's world, but it was also liberating. I liked no one hovering over me and there were no hoops to jump through in order to get my work done.

This early attitude towards safety, which was akin to throwing someone in a pool and telling them to swim, bred a culture of independent risk-taking field geologists. My first geology job was working for the USGS in Wyoming. I was working for Mac, my graduate advisor. He had hired me from California and told me to meet him in Ft. Collins, Colorado. My father offered to drive me to Ft. Collins because I didn't own a vehicle. When I got to Ft. Collins, there was a note and a map from Mac. The note said to meet him at a campground in Wyoming. My father was kind enough to drive me to the campground. I spent 6 weeks with Mac, carrying his rock samples and learning to map from one of the best field geologists I have ever met. After 6 weeks, he assigned me a Master's project, drove me to another campground and dropped me off. He went off to work in Montana. After a number of weeks working on foot from the campground, another CSU graduate student dropped by to say hello. He asked me if I had a vehicle. I said No. He then asked me how I was going to get back to Ft. Collins for school. I said Mac was going to pick me up. I was then informed that Mac wasn't going to be at CSU when school started. I was dumbfounded and despondent until the student offered to pick me up on his way back to Ft. Collins. That summer showed me field geologists were a pretty independent bunch and I had better be able to look after myself. I eventually got a vehicle and spent summers, alone mapping in the Wyoming mountains. Needless to say, some geologists I knew who were thrown into the pool didn't take to the water. These non-swimmers usually ended up doing other kinds of jobs, sometimes related to geology and sometimes not.

The safety attitude and culture in the workplace took decades to change. Over 40 years, I went from the person who just wanted to get the job done and not get killed or maimed to one who oversaw safety programs. Eventually, I became the BOM representative for aircraft safety, the head of the safety team for the BOM and taught classes in Mine Safety to BLM and Forest Service employees. Although safety is important and many employers extol the mantra Safety is Job 1, it is not a correct statement. After taking many many safety classes, I learned the number 1 priority at work is not Safety. The number 1 priority is getting the job done. Then the next priority is to get the job done safely. If a person thought about it, if Safety was the number 1 job, then no one would do anything because there is danger in everything a person does, from laying in bed too long (e.g., bed sores, muscle atrophy) to going to and from work (e.g., traffic accidents).

I attended my first safety training class as a geologist while I was working for BLM in 1978. It was an Arctic Survival course run by the military in Fairbanks. BLM thought a winter survival course was necessary for personnel living in remote duty stations. The winters in Tok are bitterly cold, with temperatures commonly reaching

30 and 40 below zero. Since BLM personnel had to drive 4 hours to get to Fairbanks or 6 hours to get to Anchorage, BLM wanted their personnel to be prepared for anything that might happen, especially during the winter months. I was very impressed by the military training, where the safety of the trainees seemed to be their main concern. During my 2 day campout in the woods in the middle of winter, with no food or water and with only a parachute and sleeping bag, there was always a trainer around in case there were any problems. This was in stark contrast with some of the private survival training courses I encountered over the years.

One year I was working out of Whittier in Prince William Sound. On an island in the Sound, I ran into some people who were taking a survival course offered by a private entity. The entity had dropped off their clients on an island in the middle of Prince William Sound, gave them kayaks and told them to get to Whittier. I ran into the trainees on a beach on an island where I was evaluating a prospect. The survivalists had landed, jumped out of their kayaks and sprinted past me into the woods looking for food. When they finally noticed me, I offered them some of my lunch, which some accepted, but others refused because it was against the rules. I then pointed out the bear sign, which they hadn't seen. The bear sign was a big cage put out by Alaska game people in order to capture an unruly bear. None of the people had any weapons, which with the preponderance of bears in the Sound, was not smart. When I got back to Whittier, the trainers were sitting in a cafe, drinking coffee. They asked me whether or not I had seen their trainees. I told them where I had encountered their charges and they thanked me and went back to drinking their coffee. I guess all of the trainees made it back to Whittier, but since I left soon after talking to the trainers, I'll never know. After seeing how some people in private industry conducted training, I decided that any future training I was in charge of would be like the military's.

After my first year with BOM, the emphasis on safety training and equipment gradually increased. Each Spring, we would spend weeks in safety training classes. The classes were coordinated by the employees. We either found experienced knowledgeable BOM employees to teach the classes or we would bring in outside experts. A favorite class where we employed outside experts was gun safety training. For the BOM, we would have the Anchorage Police Department (APD) not only talk to us about gun safety, but they would take us to the shooting range and run us through some shooting exercises. I remember one such training session given by the Armorer for the Anchorage PD. He set up people targets for us to shoot. We got a shotgun and then we would jog past the targets while he yelled head, we'd shoot at the head, body, we'd shoot at the body, body, head etcetera until we came to the end of the row. He was trying to simulate the adrenalin that was going to be pumping through our bodies if we confronted an angry charging animal.

Each year we went through mine safety training. Mine Safety and Health Administration (MSHA) instructors would come to the office and give training sessions. They would give lessons on hazards associated with mine and mill sites and how to use safety equipment. Every year the MSHA training was different. We tried to keep the training sessions from being too repetitive because after watching the same training videos for a number of years, people would get kind of numb. The best mine training involved actually going to a mine site. One year we took Tom Pittman from the BOM Juneau Office up to Crow

Pass out of Anchorage where there are some abandoned underground gold mines. Tom Pittman had worked as a mining engineer throughout the world and was a knowledge resource everyone in the BOM tapped at one time or another. While looking at some of the mine openings in the side of the mountain, we asked him if he would enter one of the openings and he said no, it looked too dangerous. We kind of looked sheepish when we admitted we had earlier entered and mapped the mine prior to receiving any safety training.

Another time, we went to the Independence Mine near Wasilla. The Independence Mine was the largest gold mine near Anchorage. A company tried to reopen the mine in the 1980s, but the bulk sample they sent to a testing lab in order to design a mill did not represent the mined rock. Therefore, after a lot of money was spent installing the mill equipment and reopening the mine, the mill couldn't recover enough gold to keep the mine going. The company, which tried to develop the mine consequently stopped all operations and leased out the mine to a couple of small miners who made a living mining high-grade portions of the gold vein. After contacting these small miners to show us their mine as part of our safety training, the miners had us crawl up and down 100-foot tall ladders and into 4-foot tall workings. Some of the mine openings were kept open because the old miners had left pillars of the quartz vein. These pillars had visible gold in them. The miners were mining the pillars and leaving behind 8-inch by 8-inch wood beams in their place. It was great fun crawling around this big old mine with experienced people.

Other safety training included aircraft and bear safety training. OAS provided aircraft safety training and if we used helicopters, we would also be trained by the helicopter pilots on-site prior to beginning work. Each helicopter pilot and each helicopter was unique, so it was important to have a safety briefing prior to starting work.

Experienced people in the office would give bear safety training. Since bears are probably the most dangerous animal encountered in Alaska, bear safety training was very lively. There are many bear stories in Alaska, and I have had many close encounters with the burly bruins. A bear encounter that was always brought up during the training was of a BLM survey crew's encounter with a grizzly. BLM made a video of interviews with the people who were involved in the incident. I would show the video to BOM field crews every year. It seemed the survey crew was returning to their helicopter at the end of the day and ran into a sow grizzly and her cub. The sow charged and mauled a male and a female crew member before another team member, who was carrying a rifle could kill the bear. He probably saved his crewmates lives; however, the woman's face was permanently scarred. Every time the video was shown, their ordeal made a very graphic and sobering impression on everyone. Years later, while working at the BLM National Training Center in Phoenix, I ran into a woman working for BLM who looked familiar. She said she had lived in Alaska. I kept on wondering where in Alaska I'd met her. Someone finally told me she was the woman who had been mauled. She looked familiar because I had seen her in the video so many times. Small world.

We never trained people in technical rock climbing techniques even though many of the BOM personnel were experienced mountain climbers. Our policy was if a person needed to do anything involving technical climbing, either up or down, then we would either contract to have the work done or not do the work. We felt we did not want to put our people into more dangerous situations than they encountered every day in the field. One aspect of training we emphasized was a person always worked to the level of the competence or comfort of the lowest employee. This sometimes ran counter to a field geologist's cultural upbringing. It was sometimes hard to "chicken-out" when presented with a challenge by a field partner.

Safety equipment also got better throughout the years. When I first started work, underground safety equipment consisted of a hardhat, carbide headlamp powered by calcium carbide and flame safety lamps used in coal mines. The carbide headlamps have a nozzle inside a reflector, a water reservoir and a reservoir for calcium carbide. Water drips onto the calcium carbide, producing acetylene gas, which comes out the nozzle. At the end of the nozzle, there is a metal roller and flint, similar to those found on a cigarette lighter, where a spark is created by rolling the metal against the flint. The spark ignites the gas and a flame is produced. The headlamps are now almost museum pieces because they aren't being made and calcium carbide is very hard to obtain. But, at the time, each geologist carried extra carbide and water. It was a chore to recharge the lamps when the calcium carbide got used up while underground. To recharge the lamps, I had to set all my field equipment down on the floor of the mine, sit down, take out the carbide and water from my pack, take the headlamp off my hard hat, unscrew the carbide reservoir, dump out the spent carbide, put in the new carbide, screw the reservoir back on the lamp, fill the lamp with water, then try to light the lamp. Many times the nozzle had to be cleaned out to get the lamp going. Working in pairs was always good because usually, only one person's lamp would go out at a time, so there was light to see while one of us performed the recharging maneuver. In 1981, the BOM replaced the carbide headlamps with battery-powered headlamps.

We also had flame safety lamps. A flame safety lamp is just an open flame in a cylinder surrounded by a screen and glass. A safety lamp's flame will burn brighter if flammable gases are present or dimmer if the oxygen level decreases. The safety lamp was developed for use in coal mines because even though the safety lamp has a flame, it will not ignite methane if it is present in the atmosphere. It took about 25 more years to replace the flame safety lamps with electronic gas monitors. But that's another story.

One of the biggest safety issues we faced in Alaska was communications. Since data showed in case of emergency the survival rate decreased rapidly over time, being able to communicate with a field crew was a big deal. When I worked for WGM, we had a single sideband radio in our field camps used to contact the outside world. I could even patch through a phone call. When in the field, there were hand-held radios used to contact the helicopter as long as it was in sight.

The BOM employed various communication methods. If I was staying at a hotel or lodge, most of the time there was a telephone available. If I was in field camps, I either had a single sideband radio or in later years, a satellite phone. Satellite phone service didn't come into common use until after 1990. To get the service required renting a satellite dish and paying a technician to set up the phone service in the field. Amazingly enough, the satellite dish we used in the Brooks Range needed to be pointed at the ground for the best reception. I guess this is the reason a technician was hired to install the equipment. For some of the smaller field projects, I would just have a schedule left with a supervisor enumerating when I needed to be picked up by an airplane or helicopter and/or when I was going to be back from the field.

When I first joined the BOM, every field crew was essentially on its own. I often wondered how long it would have taken for the people in the office to realize a field crew was not coming back. This is still a common problem with field-going crews, not only in Alaska but all over the US. With the increasing emphasis on safety, I implemented a plan for the field-going crews. Each crew had to have a communications plan, which would state where they were going when they would be back and what were the check-in procedures. If crews were staying in hotels, lodges or had a satellite phone, I wanted to know every day when the crews were safely back in camp. Every crew had my home phone number and I expected a nightly call. They also knew if I didn't receive their call, then I would call them and/or start emergency procedures. Prior to the implementation of this protocol, emergency procedures were only started once a "loved one" informed the office their husband, wife, boyfriend, girlfriend, was overdue. This new policy was met with some resistance. Why would there be resistance by field personnel when all anyone was trying to do is save their lives? I think it's the field geologist's macho attitude/culture. Many of them worked in the field alone when they worked on their college projects, private industry or on past field projects. They wondered what the big deal was with the new big brother act. Many of them didn't realize management was actually concerned for their safety because we weren't just working with colleagues they were our friends.

Two instances emphasize this point. One happened at a BOM office in the lower 48. A wife called her husband's supervisor asking if he had heard anything from her husband or his co-worker who were doing fieldwork in the Colorado/Utah area. She hadn't heard from her husband, who usually called her every night. The supervisor assured her everything was probably OK. After a week of daily calls from the wife, the supervisor sent out a crew to look for the missing men. They found them dead in the field. They surmised a lightning storm had come up and the men started walking back to their truck when a bolt of lightning hit both of them and killed them instantly. It wouldn't have mattered if the "rescue" crew would have been sent out earlier because both of the men were dead. However, this incident didn't send a reassuring message to the rest of the employees that management cared about their well-being. Big changes occurred in that office after this accident.

The second instance involved a call from Joe Kurtak one evening. Joe, who was leading a field crew in the Valdez Creek Mining District, was checking in as per the communications policy. He started the conversation by saying, "We got everyone out OK." I said, "Could you elaborate?" He proceeded to tell me one of the field crews was examining an underground working when the front of the working collapsed, trapping two people underground. This crew was on loan from the BOM Spokane office and very experienced. Joe explained the crew thought the opening didn't look good but was stable enough to enter. Even though they thought everything was going to be OK, they hedged their bet by having a third crew member stay on the surface. The third person spent his time mapping the surface geology. After the third person completed his work, he returned to the opening and found the opening had closed shut. He called the helicopter, who informed Joe. Joe mobilized the other members of his crew. They brought shovels and were able to clear the opening in a couple of hours, freeing the trapped men. The only suffering the trapped men had endured was missing their dinner.

So, why was this incident so important to me? Well, first I was glad I learned about it almost immediately and nothing bad had happened. Second, it didn't seem like something I would report to anyone other than my immediate supervisors because no one got hurt. However, it took on more importance when I got contacted by the BOM Safety people in Washington, DC about this incident. I had naively thought what happens in Alaska stays in Alaska, but didn't count on the guys from the Spokane office telling Alaska stories that would get back to Washington, DC. The Safety people wanted to know if an Accident Report had been filed. I was able to tell them the incident had been reported to me right after it happened, but I considered it an "incident" rather than an "accident", so I didn't think it was necessary to file a formal report. The Safety people informed me I was wrong and chewed on me for not reporting. My only consolation is I knew about the "incident" right after it happened and not weeks or months later. Another consolation is because I was able to communicate with Joe, if Joe needed more help, I would have been able to provide him help. The amazing thing was when a BOM field crew was walking past the underground working the next year, a head popped out of the opening. A consultant, working alone, had just mapped and sampled the working. Our crew told him about what happened just the year before. Who knows how long before anyone would have come looking for him if he had been the one who was trapped?

Field crew communications were and are a constant issue. Hand-held radios are line-of-sight. Line-of-sight means if there was a mountain between two people, then they usually couldn't communicate with each other. A way around the line-of-sight problem was the installation of repeaters. Not being a radio technician, I really don't know the ins and outs of repeaters except to say they kind of act like cell towers. If a person is in sight of a repeater, then the repeater picks up the radio call and relays it. Each repeater was set to a frequency. Purchasing, installing and maintaining repeaters for my projects was always too complicated and expensive for a small agency like the BOM, so it was nice when we could use someone else's repeater system. This rarely happened because no agency had a widespread repeater system in Alaska and many agencies were reluctant to have another agency chattering over their systems. So, for the most part, the only communicating done while I was in Alaska between field crews was when we were in sight of each other.

Even with the current technology, communicating with field crews is still an issue. Because of their ubiquitous nature, cell phones, satellite phones, I- or Android phones have taken the place of radios. However, most people learn, once they are not in an urban environment cell coverage is spotty at best. Therefore, a field geologist who goes into the boonies, has to use the age-old method of communications, which is to tell someone who cares (e.g., spouse, partner, friend, boss) when they will contact them, which may be in hours, days or weeks.

It makes sense to have a clearly defined safety policy and provide safety equipment and training to all personnel. Overall the safety equipment the BOM provided in Alaska was great. It consisted of the aforementioned flight helmets and nomex clothing, backpacks, first aid kits, hard hats, lights, self-rescuer (for underground work), gloves, safety glasses, earplugs, guns (either a .44 caliber pistol or a shotgun), radios, bear spray, mosquito repellent, head nets, field vests, with all of the accoutrements (pens, pencils, markers, rulers, measuring tape, flagging), crampons when working on ice fields, ATV helmets, life vests, ropes and all of the field equipment (e.g., hammers, chisels, sample bags, sluice boxes, shovels). The only things not provided were personal clothing, cold weather gear, rain gear, boots, toiletries and other personal gear. Boots were not supplied at the time, but when I transferred to regulatory agencies, I successfully argued geologists should be provided with steel-toed boots because they are required by MSHA to be worn on any mine site and a geologist can't do their job without them.

Mary and the author enjoying spending the winter together in Anchorage

Chapter 28

Home Life and Age

Momma's don't let your babies grow up to be geologists. If Ed and Patsy Bruce, who wrote the song, knew geologists they might have changed the protagonist of their song from cowboy to geologist. One of the lyrics of their song seems to especially fit a geologist, "Cos' they'll never stay home and they're always alone. Even with someone they love." Throughout my career fellow geologists would introduce their spouses, who struck me as wonderful people. Invariably, years later I would discover the presumed wedded bliss was no more. My geologist friends were usually still happily pursuing their careers but were not accompanied by the same partners. This phenomenon was in line with the perception the geology profession has the highest divorce rate and also the highest job satisfaction rate of any profession. Since I witnessed this phenomenon numerous times, I believe the perception is true. Fortunately, my life has been the outlier because I have been married for over 40 years to Mary. This may have more to do with Mary's forbearance than any effort on my part. It could also be the old adage "Absence makes the heart grow fonder" is true. Or it could be turnabout is fair play.

In my 20s and early 30s, I was on field projects where I would be gone 60 days or longer in the summer. Luckily, Mary either had a similar job or had a job where she could "hold down the fort" at home. If the field camps were accessible by vehicle, Mary would sometimes visit during the field season, which was a special treat. If the camp wasn't accessible by road, then Mary and I would communicate occasionally via radio or satellite phone.

Using the radio phone was always tricky because everyone in Alaska was listening. It was the ultimate party line and the source of all kinds of entertainment for bush people. Each party using the phone had to be careful about what they said to each other, so there wasn't much mushiness communicated between parties.

During the time I was gone, Mary shouldered the responsibilities of work, bill paying and home maintenance. She could try to ask my advice, but essentially she was on her own if appliances broke or anything had to be done around the house. In later years, as a supervisor who oversaw field projects, I could schedule my own field time to work around Mary's schedule. This freed her up to take field jobs, while I stayed at home. In those later years, I found out what she had been putting up with during my absences. I also became the one who would go visit her in field camps, like when she was working out of Kaktovik on the Beaufort Sea.

I always appreciated coming home after long absences and reconnecting with Mary. It was a joyous feeling walking in the front door and being greeted by a smiling face. It was also nice to come back from the field and see a face not covered with 3 days of beard stubble and who didn't smell like old sweat socks. The bad thing about long absences was the "honey-do" list waiting for me when I got home. There was always a little bit of summer left, so I had to take over some duties like yard work, outside home improvement projects and getting the house ready for winter. I usually worked on the outside "honey-do" list until the snow flew in mid-September. After the termination dust coated the mountains, it was time to tackle the items on the inside list. After working around the house and the office for a period of time, I started longing for the days where I would wake up in the morning, put on my field gear, go to breakfast, then tell the chauffeur to take me into the wilderness, walk all day, return to a prepared dinner, do office work and play until bedtime.

One advantage and disadvantage to living in Alaska was there were about 8 months of winter. This meant I worked like crazy in the summer, collecting reams of data, then I had 8 months to analyze the data and write reports. It also gave a person time to reconnect with friends and an opportunity to take long vacations out of Alaska. Most of my friends in Alaska were either gone in the summer or were so busy working, playing and fishing that there was little time for socializing. Therefore, winter was a special time when Mary and I would reconnect with friends and have pot-lucks, parties, go out to dinner and have game and movies nights. Since many of our friends did not have family in Alaska, our friends became our families. After all of these years, we still consider them our family.

Traveling out of Alaska in the winter time was a common occurrence. Almost every geologist I knew would be gone for weeks or months in the winter. Since it cost about as much to fly to Europe or Asia as it did to fly to the lower 48, I would run into Alaskans all over the world, but most commonly in warm climates. Mary and I visited New Zealand one year and chose to only go to the North Island. We didn't care to visit the famous glaciers on the South Island because we'd seen enough ice and snow. We stayed around the hot springs and beaches on the North Island, then flew to Samoa to visit family and laze on its sunny beaches. We also liked to get out of Alaska in February or March because when we got back, even though old man winter still held Alaska in his icy grip, there were a lot more daylight hours than when we left.

Geology fieldwork is very physically demanding. In my 20s, it was easy to get into shape. I would just go into the field and after the first two weeks I could hike all day and when I got back to camp, I could do all of the paperwork and then play games, like volleyball, darts or cards. When I reached my 30s I spent the winter in the gym trying to keep in shape because I knew I couldn't rely on those first two weeks in the field to get me into the shape I needed to be in to accomplish the needed work. The problem with Alaska is the winters are cold and dark and it was hard to get motivated to go from work to a gym. This was especially true when all I really wanted to do was to navigate the icy streets and get home safely. But, it had to be done.

Age seemed to catch up to me in my late 30s. I was working in the Brooks Range and after I would get back to camp I thought, "Hey, I'm still feeling good after a day's work." It's then I realized I unconsciously was having a "Better Life Through Chemistry." When I got back from the field, I would start drinking coffee and drink it throughout dinner. I think the intake of a lot of caffeine was the only way I kept from collapsing at the end of the field day. It's also a time when many geologists become familiar with Mr. Arthur Itis. This guy, more commonly known as arthritis, takes up lodging in a geologist's hands, knees and spine. Swinging a hammer onto mostly unyielding rock for many hours takes its toll on the joints in the hands. Carrying heavy packs and lugging thousands of pounds of equipment and rocks over the years is especially hard on knees and spines.

In my late 30s, I started having trouble with my knees and found out I had torn the meniscus in my right knee during a traverse in the Brooks Range. When I got back to Anchorage, I saw an orthopedic surgeon who said I could have surgery or just live with occasional pain. Since I haven't had too much luck with any kind of surgery, I decided to live with the pain until it becomes unbearable.

Every person is physically different, but it seems geologists gravitate towards limited fieldwork as they age. When I was doing fieldwork in my 20s, the big bosses were old men in their late 30s and 40s. They would sit around and tell stories about going into the field. If they felt like it the bosses would come out to visit the field camps to see how I was doing or to just relive their glory days. They would fly in, spend a couple of days going out to field sites to see what we had found. They were usually not very mobile. Then, they would fly out and I would report to them at the end of the season. I kind of felt sorry for these old codgers. Little did I know the wear and tear of the field on my body would eventually make me gravitate towards that kind of job. Therefore, when I was offered a supervisor job in my late-30s it was a tough decision. I would be giving up a life in the field for one of meetings, budgets and personnel. After much soul-searching, I took the job, which eventually morphed into supervising all BOM research projects in mainland Alaska. I would still go into the field, but on a limited basis, usually only for two week periods.

I know many geologists who can still do fieldwork and collect their Social Security checks, but usually, it's for a limited time and duration. Instead of going out for two months at a time, a senior geologist goes out for two weeks or two days. There are, however, exceptions. I know geologists in their 70s who can outwork a lot of younger people.

Tony Dunn and a contract pilot on top of Slug Mountain, southwest Alaska, two weeks before he was killed in an accident

Chapter 29

Death in Alaska

Alaska seems to have more than its share of untimely deaths. I refer to them as untimely because many people I knew didn't get the chance to lie on their deathbeds after living a full life surrounded by their loved ones. Since I knew a lot of geologists and mining people I was extremely sensitive to their demises. Geologists in Alaska are mostly young and adventurous. They work in hazardous conditions where one mistake can cost them their lives. My first exposure to death was in 1977 when I was working for WGM. We were camped near Kokrines on the Yukon River. We got a radio call from one of the WGM camps. There had been a plane crash and one of the seasonal geologists had been killed. The camp where the tragedy happened was being moved using a plane on floats. On one of the runs, the plane crashed on takeoff, and the WGM seasonal employee, who was a passenger, was killed. They needed the use of our helicopter to help transport the body.

The second exposure to death occurred when I moved to Tok in 1978. I was told a miner was murdered on the Fortymile River just before I arrived in February. Two miners had a dispute with another miner over the same portion of the Fortymile River. The two miners ambushed the third miner and shot him numerous times. The shot miner lived long enough to tell his story to the Alaska State Troopers. The perpetrators were caught and tried that summer. In a twist, the jury convicted one of the ambushers and let the other one go, even though two different bullets were found in the body. Needless to say, the locals, which included me and those living and mining along the Fortymile River were not happy this accused murderer might come back into the country. The story had a fitting end when the ambusher was found dead a couple of years later. He died after he returned from a trip to Asia. His death was caused when a balloon full of drugs he had swallowed in order to smuggle them into the US ruptured, which caused a fatal overdose.

One of my co-workers, Brian Zelenka, died while taking a mountain climbing class. He and his climbing partner were taking the class to prepare themselves to climb Mt. Foraker, the mountain next to Mt. Denali. On one particular day, the climbing class was held on Pioneer Peak near Palmer. After the class was over for the day, Brian and his climbing partner decided to climb to the summit of Pioneer Peak. According to Brian's climbing partner, they were coming down from the summit and hit a patch of snow/ice. The climbing partner was able to stop his descent, but Brian couldn't arrest his slide and fell 1,000 feet to his death. Brian was one of those enthusiastic people who was fun to have in the office and the field. He loved being a geologist. The death of Brian didn't deter his climbing partner from attempting the summer climb of Mt. Foraker. He found another climbing partner who wanted to climb Mt. Foraker. Unfortunately, during the climb both were killed by an avalanche.

Tony Dunn, one of the BOMs' seasonal employees died in 1986. Tony was an experienced geologist who had been laid off by a mining company. He had been caught in one of those mineral industry wave troughs, which so many geologists experience. He worked seasonal geology jobs and during his time between jobs, organized the SLUGS (Society of Laid Back Unemployed Geologists). I hired him for my project at Goodnews Bay, which only lasted a month. One of the highlights of his work with me was he got a picture of himself on top of Slug Mountain in southwest Alaska. After working for me, he transferred to the Juneau Mining District project, which surprisingly was based out of Juneau. Southeast Alaska is mostly mountains rising from the ocean cut by steeply incised creeks. Work in Southeast Alaska involves either climbing up or down. It was on one of those climbs where Tony's luck ran out. He was working with a permanent employee. They were climbing up a mountain and Tony was having a hard time. During the climb, he would get the shaky leg syndrome, which happens to a lot of people who find themselves in a precarious position. I call it the shaky leg syndrome because when a person finds themselves in a position, like on a cliff face, where they think they might fall, their legs can start shaking uncontrollably. Eventually, control will return if a person is assured of not falling. This happened with Tony. The permanent employee explained to Tony he would look after him and it was safer to go up than to try to go down. The employee told Tony when they reached a flat area, he would go back to town and get a helicopter to get Tony off the mountain. The employee had to physically go for help because they couldn't communicate with the office. They had reasoned, wrongly it seems since they were working so close to town, setting up communications prior to climbing the mountain wasn't necessary. They followed the plan as outlined. The employee helped Tony to a safe place where a helicopter could land and he went for help. When the employee returned with the helicopter, he found Tony's body at the bottom of a snow slide. It was surmised Tony, having overcome his shaky legs, decided to cross a snowfield. No one knew why. It seemed he had slipped and slid headfirst down the snow slide onto some rocks at the bottom of the slide.

Over the years it seemed death of people my wife and I knew in Alaska was a constant. One winter a geologist I knew died in an avalanche while backcountry skiing near Anchorage. Ralph Yetka, who I mentioned earlier, died in a helicopter crash along with a USGS geologist. Dennis Southworth, who worked on and off for the BOM over the years and who I worked with on various projects, died an early death. Joe Vogler, the Libertarian firebrand and miner, who everyone in Alaska knew and with whom I was acquainted, was murdered at his home in Fairbanks. My wife worked for and with four people who died. One of her bosses was murdered by his son and one boss died in a plane crash. One of her former co-workers was murdered by an escaped convict while camping with her boyfriend in the lower 48. Another co-worker, a polar bear biologist along with another biologist and a pilot disappeared while conducting aerial polar bear surveys north of Point Barrow. To our knowledge, they were never found. Ex-Senator Ted Stevens, who I met a number of times, was killed in a plane crash years after surviving a plane crash which killed his wife. Lately, a former co-worker of mine died when she and her husband took a flight-seeing trip associated with an

Alaskan cruise. Since death seemed to be a common occurrence in Alaska, I never thought much about it, except
to try to help the odds by training, planning and vigi-
lance.

I knew Alaska was unique after I moved to Arizona.
Only one person I knew in Arizona had experienced a
death of someone they knew. That person, on my mention of having lived in Alaska, told me his sister and
her husband were murdered in Fairbanks. Everyone else
I knew eventually experienced a death of friends and
loved ones, but their friends and loved ones died "natural" deaths surrounded by family members.

Kennicott Mill, Wrangell-St. Elias National Monument

Chapter 30

Afterward—Alaska Then and Now

When I started working in Alaska in the 1970s, Alaska was changing, again. The days of unfettered access to all of the State were ending as were the lives of some of the people who had been involved with the early development of Alaska. Prior to Statehood, in 1959, most of Alaska was owned and managed by the Federal government. People did not have to worry whose land they were on because there was a good likelihood whatever they were doing, they were doing it on Federal land. When Alaska was admitted into the U.S. as a State, it was allowed to select 90 million acres out of the approximately 365 million acres of available land. To make sure it selected the best lands, the new state government set up a way to prioritize its selections. I don't know the various criteria other than the State wanted to select land around its cities, like Anchorage, Fairbanks and Juneau and it wanted land that would provide an economic benefit to its citizens, like land with oil under it as in the case of Prudhoe Bay. Even though from Statehood until the early 1970s, the State of Alaska evaluated and selected high-priority lands, they didn't seem to be in any hurry to get all of the selections made. This all changed with the discovery of the vast Prudhoe Bay oil field, which is on State land. In order to develop the field and build the pipeline necessary to move the oil to market, the State and Federal governments first had to deal with the land claims of the Indigenous or Aboriginal people who lived in Alaska. This issue had been brewing for a long time but the development of the Prudhoe Bay oil field finally brought it to a head.

In 1971, the oil pipeline was given the go-ahead when the Federal government passed the Alaska Native Claims Settlement Act (ANCSA), which formed Native corporations and allowed those corporations to select 44 million acres of land. Generally, Native selections were given priority over State land selections, which put the State selections on hold. The Native selection criteria were similar to the State selection criteria (i.e., selections around villages and land that could provide an economic benefit to their shareholders).

Additional provisions in ANCSA withdrew vast acreages of Alaska from selection by either the State or native corporations. Many of these areas were eventually designated as National Parks, National Monuments, Wildlife Refuges, National Forests and Wilderness areas with the passage of the Alaska National Interest Lands Conservation Act of 1980 (ANILCA). When all was said and done in 10 short years Alaska changed from a land where almost 100% was open to mineral exploration, to one where an area about the size of California was withdrawn from mineral entry. This doesn't account for the restrictions associated with Native and municipal lands, which combined with the withdrawn Federal lands equals an area about the size of Texas.

Not only did the passages of ANCSA and ANILCA withdraw millions of acres from mineral location, but the passage of various laws in the late 1960s and 1970s made mineral development harder in those areas still open for claim staking. These included the Wilderness Act of 1964, the National Environmental Policy Act of 1968 (NEPA), the Clean Water Air Act of 1970, the Clean Water Act of 1977, and the Federal Land and Policy Act of 1976 (FLPMA) to name a few. The regulations associated with NEPA, clean air and water and FLPMA have made permitting a mine a huge and costly process. The environmental documents associated with mine permitting, which in the 1970s and early 1980s started out as small books have grown into such huge documents that they take up whole bookshelves.

There has also been a shift in attitudes in the last half-century. There have always been fringe ecological groups who abhorred the thought of any kind of development. However, over the years, the ideals of these groups have become mainstream. In Alaska, when ANILCA was passed, President Carter was burned in effigy and there was a serious movement to secede from the U.S. Now, most people have gotten used to the idea of preserving wild places and could not imagine a country where mineral companies would act like they did during the industrial revolution. I think most people approve of governments requiring mineral companies to develop resources in ecologically responsible ways. Over the years I have noticed a growing number of Alaskans who have also developed the NIMBY attitude toward mineral development. NIMBY stands for Not in My Back Yard.

I witnessed many changes due to new laws, regulations and attitudes over my 18 years in Alaska. These changes, good or bad, created a lot of work for this field geologist. When I came to Alaska in 1977, WGM was evaluating land for possible selection by the Doyon Corporation. I played a very small part in adding information that led to their final land selections.

In 1978, when I went to work for BLM in the central uplands and lowlands system of eastern Alaska, the new Act (FLPMA), which gave BLM a framework to manage Federal lands was little over a year old. I was on the front lines trying to make sense of the new law, which has clear wording such as not letting miners cause undue or unnecessary disturbances. Some of the miners I knew argued since the creeks they were mining flowed into the Yukon River and the Yukon River was so dirty that when one was floating on it the sound of dirt scratching against the boat could be heard, it wasn't necessary to build settling ponds to stop their dirty water from proceeding downstream. BLM didn't think much of this argument and I had to try to convince miners that mining in an environmentally sound manner was good. It took about 10 years for BLM to clarify many of the provisions of FLPMA via regulations and many years after that for the miners to figure out what was due and necessary.

I also had to deal with the fallout from the passage of ANCSA and eventually ANILCA while working for BLM. The Fortymile Resource Area was millions of acres in size, but because of ANCSA, there were portions withdrawn from mineral entry and portions of the area's land status that had to be determined by the government. These areas included the Fortymile River, where mining had been conducted for over 80 years. Because of the mineral withdrawals, I had to make sure no one staked claims or mined in closed areas.

I also worked on determining navigable waters and

historic trails in the area. A navigability determination was important. If a body of water was determined to be navigable the State would own the river or creek bottom. The land the State acquired by such a determination was not deducted from the 90 million acres the State received at Statehood. If the body of water was non-navigable then the landowner would own the bottom. I don't want to get into the legalities of navigability, but I'll just say the State wanted as many bodies of water to be navigable as possible and landowners and managers like the Native Corporations, National Park Service and Fish and Wildlife Service wanted as many bodies of water to be non-navigable. As a result of BLM's navigability surveys, one of the rivers, the Fortymile River, which was designated as a Wild and Scenic River, was deemed navigable. This was a big victory for suction dredgers on the river because prior to the navigability designation the river was off-limits to any new mining claims. After the designation, suction dredgers can stake State mining claims on the bottom of the river and mine those claims.

I also worked on determining if there were any historic trails or rights-of-ways in the Fortymile Resource area. A historic trail or right-of-way is one that has been in existence prior to the enactment of FLPMA in 1976. Prior to FLPMA, a "highway" to use the term loosely, could be granted across any public land without being formally recorded. This was known as Revised Statute 2477. FLPMA repealed this statute. Since 1976, BLM has been dealing with the question of whether or not an existing road or trail is actually a right-of-way. This has led to a lot of conflicts not only in Alaska but throughout the west. The designation of rights-of-ways for the State of Alaska was important because a large amount of land was selected by various entities along highway corridors.

Therefore, the State wanted to maintain as much access from those highway corridors as possible. Of course, the new landowners or managers along those corridors did not want to let people have unrestricted access across their lands. To determine whether or not a trail or road qualified I researched historical documents, maps, aerial photos and performed fieldwork to see if a trail actually existed.

FLPMA required BLM to develop land management plans. A land management plan inventories and evaluates all of the resources in an area. These plans are used to guide agency decisions and identify use conflicts. For the Fortymile Resource Area's plan, I had to inventory and evaluate the mineral potential of the area. I also had to determine the likelihood of future mineral development, with little or no fieldwork. This exercise gave me a good understanding of the Federal government's use of WAG, which stands for Wild Ass Guess. Here I was a 26-year old, who had worked in an area less than a year, determining which parts had the highest mineral potential and what the metal prices were going to be in the next 5 years. I did the best I could with my limited knowledge and experience, but I don't think I did very well. I definitely WAG'd it.

My last assignment with BLM was examining gravel deposits along the proposed gas pipeline route. In 1979, oil was flowing through the pipeline from Prudhoe Bay to Valdez. However, the gas produced is reinjected underground because there is no way to get it to market. Therefore, there was a proposal to build a gas pipeline parallel to the Alaska Highway and move the gas through Canada to the lower 48. I guess this is another example of how Alaska has changed. The oil pipeline was built in less than 10 years. However, even though there have been numerous studies and changes to the gas pipeline routes, it is still on the drawing board as I write.

The final result of Statehood and the passages of AN-CSA and ANILCA resulted in most of the land in the BLM Fortymile Resource Area being either designated a National Monument (Yukon-Charley), a Wild and Scenic River (Fortymile), Native Corporation land or State land. Because the only land BLM now manages is within the Wild and Scenic River corridor, the BLM office in Tok was closed and all of the staff reassigned. Luckily I missed all of that fun.

I continued to be positively and negatively impacted by the changes occurring in Alaska after I left BLM for the BOM in Anchorage. Prior to 1970 and the passage of many of the aforementioned laws, the BOM in Alaska was mostly involved with petroleum and mine safety. My boss, Don Blasko was the last petroleum engineer with the BOM. He somehow survived with the BOM after the oil functions were transferred to another agency. After 1970, the work done by the BOM in Alaska evolved into conducting mineral inventories and evaluations. The agency was uniquely qualified for this kind of work. It was organized into three offices. The main office in Juneau coordinated with the State government and had the best mining library in the State. The office in Fairbanks coordinated with the University of Alaska, Fairbanks and ADGGS. The office in Anchorage, where I worked, coordinated with the various Anchorage-based Federal and State agencies. Each BOM office had geologists and engineers. The administrative and support staffs were shared between offices. Projects were assigned to whichever office could handle the work. The field offices also coordinated research projects with the BOM research centers located throughout the U.S.

When I joined the BOM in 1980, it had just finished its work dealing with ANCSA-related issues and had started on projects created by the Wilderness Act of 1964 and subsequent lawsuits. The Wilderness Act affected US Forest Service lands in Alaska, which are located in the Pacific mountain system on the Kenai Peninsula, in Prince William Sound and in Southeast Alaska. These were the RARE II studies I mentioned earlier in the book. I worked on the RARE II study of the Chugach National Forest from 1980-1983.

The passage of ANILCA created and led to the expansion of parks and monuments. Denali National Park and Preserve's boundaries were expanded to the west and southeast. The park encompassed the traditional mining areas of Kantishna on the west side of the park and Dunkle on the southeast side of the park. The National Park Service didn't want mining in these areas; therefore, they hired a consultant to determine the value of the existing mines and mining claims, so they could condemn the claims and mines and compensate the owners. After reviewing the report the miners thought the values the consultant came up with were way too low and the Park Service had unfairly influenced the findings in the report. Therefore, the miners appealed to their Congressional delegation in order to get another study done by an impartial party. The Congressional delegation obtained funding for a new study by an unbiased third party (i.e., the BOM). The job was too big for the BOM to handle; therefore, the BOM hired a number of consultants, which consisted of Salisbury and Dietz out of Spokane, and Chuck Hawley and WGM from Anchorage. The BOM crew, which included Joe Kurtak, Bob Hoekzema, a seasonal employee and I were directly involved in conducting portions of the fieldwork. My boss Jake oversaw the contract and Nathan Rathbun provided logistical support. Bob and I spent a couple of months in the area evaluating the creeks to determine their placer gold potential, while Joe helped evaluate the hard rock mining potential. It was a great summer, with Nathan and I winning the 4th of July horseshoe pitching contest. I think we won because we were the only sober participants. Nathan and I remained non-inebriated because we had to take some supplies to Anchorage and the Park Service would only let us drive the Denali Park road after 9:30 p.m. Driving the Denali Park road when there was no traffic was one of the highlights of the summer. If the weather was clear, I would take my time and just soak in the magnificence of Denali glowing at dusk, caribou wandering over the hills, grizzly bears foraging in the tundra and mountain sheep lazing along the mountainsides. No tourists, no traffic, what could be better.

After the field data for the Kantishna study was collected, samples analyzed and economic evaluations completed, the report was finalized and presented to the Park Service and the miners. Since BOM was a disinterested 3rd party, the BOM moved onto its next projects after delivering the report. The net result was miners were compensated and mining was eliminated from the park. Now, instead of mining gold in the Kantishna area, there are lodges who are mining tourists and charging them over $500 per night. A person who wants to get a feel of how a miner's life was like in the late 1900s can book a gold-panning excursion during their stay in Kantishna.

The passage of ANILCA led to more work for the BOM. Title X, Section 1010 of ANILCA gives the Secretary of Interior the authority to expand the database of oil, gas and minerals as well as assess the mineral potential of all public lands in Alaska. The BOM used this authority to acquire funding to conduct mining district studies in Alaska. There are 67 mining districts in Alaska. After consulting all of the land management agencies, the BOM prioritized the mining districts based on available data, land status, the need for additional data, national needs for the contained commodities and physical accessibility of the districts.

In order to determine which mining districts should have priority, the BOM spent 1984 doing reconnaissance work in various mining districts. I worked on the Yentna Mining District which is located in south-central Alaska. Because of limited funding, I accessed the district by using 4-wheel drive vehicles, boats and a small amount of helicopter time. I and whomever I could convince to go with me would just camp out in a tent, and provide our own sleeping bags and food.

During my reconnaissance work, Joe Kurtak and I had a memorable experience in the Talkeetna Mountains of the Yentna Mining District. I had looked at the maps and literature of the Talkeetna Mountains and decided there was a concentration of mineral deposits in the Iron Creek drainage that could be examined with less than a week of fieldwork. However, the only ways to get to them would be to backpack in or use a helicopter. Since I really didn't want to spend a couple of weeks backpacking into the area, then trying to pack hundreds of pounds of rock samples out, I rented a helicopter out of Wasilla. I had the pilot fly Joe, me and all of our field gear strapped to baskets on the helicopter's skids to a cirque overlooking Iron Creek where we set up camp. The campsite was ideal. A cirque is a bowl carved out by a glacier near the crest of a mountain. The site had a nice clean stream running through it, a little meadow where we could pitch a tent and some of the mineral sites were at the head of the cirque. The site was also high enough to have very few bugs. We spent 5 days at the site. After mapping and sampling the close sites, we would get up early, eat breakfast and hike down into Iron Creek, which was located about 1500 vertical feet below our camp. We sampled prospects along Iron Creek and stashed those samples in a flat area on the creek. We prospected the gulches on the north side of Iron Creek and found some unreported zinc occurrences. We also examined prospects on a side drainage to Iron Creek. This required us to climb through a mountain pass 2500 feet above creek level and walk down a couple of thousand vertical feet to the prospects. We took a couple hundred pounds of samples from these prospects and stashed those in a flat area near the prospects. We then had to climb back over the mountain and up 1500 feet to our campsite. Luckily, we were both in our early 30s, Joe ran marathons and I, well let's just say I eventually made it back into camp. The only bad thing was one day when I examined my pack, I found I had been carrying around 10s of pounds of rock samples that somehow got overlooked in the bottom of my pack. Hopefully, I was just forgetful and hadn't been pranked. I didn't see Joe snickering, so I guess I just messed up. Camp was always nice to come back to because while making supper we could watch the mountains turn a rosy hue when the sun dipped below the horizon and rainbows form after a shower. Every night I was reminded of my boss's expression, which was "All this and money too." When the helicopter picked us up at the end of our stay, we retrieved the samples we had stashed throughout the area.

The biggest mining area in the Yentna Mining District was the Peters Creek/Cache Creek area. It is accessible via the Petersville Road, so evaluating the area was cheap. All I needed was to find someone to go with me, put together camping gear, buy food and pack the truck. Since all mining in the area was for placer gold, I packed the placer sampling gear. It was a nice project because we were able to go into the area, look around, do some sampling, then go back to the office when the food ran out. After evaluating the samples, I was able to go back into the area and follow-up on the previous work. The whole area was covered with mining claims so camping would have been a problem if it hadn't been for a miner I met during the first trip. His name was Martin Herzog and while talking with him, I learned his daughter, Denise, was studying to be a mining engineer. After talking with him about mining, he let me set up camp on his claim. Years later, the BOM hired Denise when she graduated from the university in Fairbanks.

I acquired a good piece of information about the source of the gold in the Peters Creek area by talking with the miners. They told me there was a gold-bearing conglomerate bed overlying the area. The gold mined in the creek beds was concentrated by the erosion of the conglomerate. Miners tried mining the conglomerate, but because it contained a lot of clay, it was impossible to mine economically until weathering broke down the clay and released the gold from the sticky substance. Studying the conglomerate was useful for the next year's study in Goodnews Bay where there was also a clay issue and subsequent studies of another conglomerate bed in the Valdez Creek Mining District.

The worst thing about the study was it was the only time someone took a shot at me. Usually one of the best aspects of working for the BOM was most of the miners liked the agency. BOM only did studies and was not regulatory; therefore, it was non-threatening. However, it seems someone didn't get the word our work was non-threatening. One day, Gary Sherman and I were taking a sample in Peters Creek when we heard a gunshot and a ricochet off some rocks near us. We looked up and saw a bunch of people with rifles on a ridge above us. Gary decided the ricochet was in line between us and the people on the ridge, which meant the shooter failed to factor in the drop of the bullet when he fired. Since we were outgunned, we decided the better part of valor was a hasty retreat or some such nonsense. We packed up our gear, got in our truck, and went back to camp. We told Mr. Herzog what happened and decided since we were almost out of food, we might as well pack up and go home. We packed up our camp, drove out and reported what happened to the nearest Alaska State Trooper. We could have confronted the shooters, but we decided the people who get paid to do such things should take the lead. I like to give the shooter the benefit of the doubt and think the guy who shot at us was either trying to warn us away from sampling someone's claim or was just fooling around. Either way, Gary and I had been in Alaska long enough to know sometimes people in the bush were in the bush because of their anti-social propensities.

In 1985, the mining district studies started ramping up. Choosing mining districts to study was now complicated by the passages of all the various laws. BOM decided the Yentna Mining District had a low priority because there wasn't enough Federal land in the district to justify a major investigation. The Juneau Mining District, in southeast Alaska, was determined to have the highest priority in the State because it contained a high proportion of Federal land; therefore, it was the first district evaluated.

BOM was a pretty lean organization. The Anchorage

office only had two full-time geologists and three full-time engineers. The Juneau office was similarly staffed. The Fairbanks office had one full-time geologist and one full-time engineer. Because there were so few full-timers, projects were supplemented with supervisors, seasonal employees and moving people around the State. If a seasonal employee was good and willing, the BOM would keep them on for the length of a project, but unfortunately, the final report was usually the end of their service with the BOM because we had such low turnover of permanent employees. Since the Juneau Mining District study was started first and therefore received all of the mining district funding, the Anchorage office sent Joe Kurtak to help the Juneau office. After spending six weeks helping the Fairbanks office look for platinum at Goodnews Bay, I was also assigned to help the Juneau office evaluate the Porcupine Mining area near Haines.

The Porcupine Mining area was a fun project mainly because the old crew from the RARE II, Kenai Peninsula study was back together. Bob Hoekzema, Nathan Rathbun, Gary Sherman and I loaded two Suburban-type vehicles with field gear and drove to Haines, then up to Porcupine Creek, which is located about 30 miles northwest of Haines. The only problem we had was going through Canada. Nathan had put together a list of all of our equipment. The Canadian customs people wouldn't take our list at face value and had us unload all of our equipment. I think they were just jerking us around because we were U.S. government employees. On Porcupine Creek, we camped out on Jo Jurgeleit's mining claim. Jo has since passed away. She was a very salty and independent woman with a wooden leg who had owned and worked her placer gold claims for years. We spent many nights having her regale us with stories of mining on Porcupine Creek. We were able to identify the placer gold resources of the area. We also discovered the gravels in Porcupine Creek were fairly thin until the mouth of Porcupine Creek was reached. Near the mouth of the creek the gravels were so thick we were unable to determine the depth to bedrock. I think they were so thick at the mouth of the creek because a glacier that had occupied the main valley had scoured out the main valley to an unknown depth, then the valley was filled in by glacial outwash and outwash from the creeks in the area. After a couple of weeks in Porcupine Creek, we drove back to Anchorage and I got ready for my own mining district study.

In 1986, I was assigned the Goodnews Bay Mining District. This district was chosen because it was pretty compact, could be completed in one year at minimal cost and since I worked in a portion of the district in 1985, I was familiar with the area. It was also low-lying and along the southwest Alaska coast. I could, therefore, start work in May before the Juneau Mining District fieldwork ramped up and use two of their seasonal employees. I completed all of the preliminary work in the winter, gathered the crew and equipment and flew out on May 19th, my wife's birthday. What a great birthday present she got.

Part of the planning for this project involved contacting the various landowners, who had to be contacted to get their permission to evaluate their land. This included the Goodnews Bay Mining Company, the Fish and Wildlife Service (USFWS), the Calista Native Corporation, the Goodnews Bay Village Corporation and the Bureau of Indian Affairs (BIA). This was the first time I encountered the NIMBY attitude in Alaska. I had no problem with the USFWS, the BIA or the Calista Native

Corporation but ran into a road-block when dealing with the Goodnews Bay Village Corporation. The Goodnews Bay villagers did not want any information about their land. They figured if people knew what was on their land, then things would change. This ran counter to my way of thinking where knowledge is good and also counter to their college-educated manager's way of thinking, but the Village council prevailed. I accepted their ruling and avoided their land. I figured they didn't need another outsider telling them what was best for them.

After dealing with the bureaucracies, I compiled all of the literature and maps; gathered all of the field gear and people; bought food; and took it to the airport for transport to Platinum. The crew took commercial flights to Platinum. We stayed at the Goodnews Bay Mining Company's camp on the Salmon River. It was a couple of huts with some beds, tables and chairs. We cooked on a portable gas grill and bathed with heated water in a 5-gallon bucket. Most of the work in the district was done using 4-wheelers. Luckily, I was able to coordinate our work with the work of the BIA who had a helicopter at the village of Platinum. The helicopter was used, when available, to access the far reaches of the district. Since the weather was usually nasty and cold, Dennis South-worth, a temporary employee, came up with the expression we used every day, "I like it here." Every morning we would don our long johns, pants, flannel shirts, fleece coats, full face masks, rain gear, gloves and rubber boots and step out into the freezing rain and say, "I like it here" as we went to shovel gravel, map, and pound on rocks. After about 3 weeks, the seasonal employees were sent to work in the Juneau Mining District and they were replaced by my supervisor Bob Hoekzema and a couple of other people who wanted to see the district.

Luckily the replacements brought some needed necessities like candy bars, crackers, potato chips, fresh fruit and10 pounds of potatoes for the last two weeks of work.

As I stated earlier, the passage of FLPMA required BLM to develop land management plans, which in the late 1970s were pretty basic. However, as the years progressed, BLM decided they needed better data on which to base their land management decisions, so they contacted the BOM to help fulfill that need. I guess BLM wanted to turn the WAGs into SWAGs. The S in SWAG stands for scientific. One specific study I was involved with was the evaluation of the mineral potential of the White Mountains near Fairbanks in 1987. The White Mountain study was a joint project with the ADGGS and USGS. The ADGGS and USGS mapped the geology and evaluated the lode potential, while the BOM evaluated the placer potential. Mike Balen and I spent a month sampling the drainages in the area. We were assisted by some geologists and engineers from the BOM Spokane office. Nathan Rathbun joined me when some of the Spokane people left the project. Placer sampling was done using a backhoe and high-banker in the areas accessible by road and shovel and mini-sluice boxes in the helicopter-accessible areas. Luckily Mike said he could operate a backhoe and I found out he could. This was back in the day if a person in Alaska said they could do something, then people would let them do it. Now, if I wanted to do the same thing, there would probably be all kinds of rules and regulations adding extra costs to a project. The data we collected was given to engineers and mineral economists who weaved it into a whole story of the mineral development potential for the area.

About this time I found that not only was Alaska changing, but so was I. After the White Mountains study was completed, my boss offered me a position as supervisor over all of the Anchorage office research projects. In order to make the decision to take a job where I would be riding a desk most of the year, Mary and I drove out to Portage Glacier, which is located about 30 miles southeast of Anchorage. It was one of those unusually sunny warm days in the Spring and we sat in the sunshine watching the glacier recede. As a wise man said, "When you come to a fork in the road, take it." So, after discussing the momentous decision with my wife, I embraced the dark side. I didn't know at the time if it was the right decision, but I knew BOM supervisors were called upon to help with field projects because we were always short-staffed; therefore, I wasn't totally abandoning my passion for the field by being a bossman.

After becoming a supervisor, I helped out in a number of the mining district studies. During the Valdez Creek Mining District study that ran from 1987 to 1990, I evaluated the portion of the Peters Creek area that was in the district. Yep, I went back to the area where someone took a shot at me. I packed up my field gear and headed out. This time I took an insurance policy with me, Martin Herzog's daughter, Denise. She had recently been hired by the BOM. We stayed on her dad's claim and set off each day walking the hills and creeks northwest of her dad's claim. It was always a joy for me to climb onto the ridges and take in the majestic Alaska Range.

I also took on the evaluation of the Tyone Creek mining area. Tyone Creek is on the extreme southern portion of the Valdez Creek Mining District. It is located about 20 miles north of the community of Eureka, which is on Highway 1, more commonly known as the Glenn Highway. Tyone Creek is a historic gold placer mining area. The mining district's project leader and I decided since the main crew was staying such a long distance from the Tyone Creek area, it made sense for me to get a couple of people and access the area via ATV's from the Glenn Highway. I made a couple of trips into the area. One was a reconnaissance trip with Mark Meyer. During that trip, we discovered the area was covered by a conglomerate bed, similar to the conglomerate found in the Peters Creek area. The best gold values found in the main drainages were where side creeks drained the conglomerate bed, just like in the Peters Creek area. We also found small amounts of platinum with the gold. I concluded the area needed a lot more fieldwork than two guys could do in a week.

The following year, I mounted a bigger effort in the Tyone Creek area. The crew was comprised of Denise Herzog, Nathan Rathbun, a seasonal employee and myself. I contacted a miner and he let us use his cabin on Yacko Creek in the Tyone Creek area. We slept in tents but used his cabin for cooking and an office. The ATVs were not only useful for getting the crew to and from the field, but they were also useful for sampling the rolling tundra covered hills. We were able to drive the ATVs on top of the ridges and sample the top of the conglomerate bed. We would dig a hole and load 5-gallon buckets with weathered conglomerate material. The buckets were then strapped on the back of the ATVs and driven to a creek where we could run the material through a portable sluice box. The only problem was the 5-gallon buckets made the ATVs top-heavy, so we had to be careful driving on side slopes or the ATV would end up on its side, spilling the material and us all over the slope. Luckily, we found when we fell off the ATVs the spongy tundra made landing OK. Not only did we discover the advantages of working with ATVs in Alaska, we also definitively proved the conglomerate bed contained gold and platinum.

In 1990, the Valdez Creek Mining District study was winding down and the BOM initiated a new and essentially the last major mining district study, the Colville Mining District (CMD). The CMD is located on the north slope of the Brooks Range in the National Petroleum Reserve-Alaska (NPRA). This district was prioritized because BLM wanted mineral data for a land use plan of NPRA. Luckily, we were able to sell the project to Congress by marrying ANILCA with FLPMA. Mark Meyer and Joe Kurtak led that effort. I spent some weeks in 1990, 1991 and 1992 helping out the project leaders, which essentially entailed just being a field grunt. Mark Meyer and Nathan Rathbun put together elaborate field camps with a satellite telephone system and propane showers. The camps were mobilized at the first of the season and demobilized at the end of the season. Since I previously related some of my experiences in the Colville Mining District, I would just say the BOM mapped and sampled 22 known mineral occurrences and discovered an additional 30 mineral occurrences during this study.

My main duty as a supervisor for BOM was to get projects funded. I would usually pitch about 10 projects a year to the Washington Office and some to Congress. I hoped one or two would be funded each year. I was told that Congress prefers to fund projects with 5 to 10-year time frames. After 5 years, if a researcher wants to remain working they had better have another project to pitch to Congress or be able to reinvent their old project with a fresh slant and the latest buzzwords. Therefore, it wasn't too surprising that after 5 years of conducting mining district studies, it was getting harder and harder to receive the funding needed to continue these studies.

Therefore, while watching the mining district studies becoming "long-in-the-tooth", I worked on getting funding for a couple of other big projects. One was a strategic and critical minerals study of Alaska. It was to be a multi-year, multi-million dollar project that got funded during the Bush administration. Strategic and critical minerals have elements required by industrialized countries. These include such things as platinum for catalytic converters, titanium that is a high strength and low weight metal, and rare earth elements used in electronics to name a few. BOM systematically evaluated all of the minerals and determined which ones the U.S. was most dependent on from foreign sources. The BOM then prioritized the list and started work. The work was to determine if there were mineral deposits with strategic and critical elements in Alaska, where they were located and how much of the element was contained in them. The studies were in cooperation with the BOM's Research Centers and the University of Alaska, Fairbanks. During the years this study was conducted, I was able to evaluate platinum deposits at Goodnews Bay, tin deposits in the White Mountains, and the gallium and germanium potential in all of Alaska. I also supervised the team of geologists, which included Jeff Foley, who systematically evaluated the potential for strategic and critical minerals in all of Alaska.

The other project that got funded was the abandoned mine inventory of Alaska. There are thousands of abandoned mines in Alaska. Many of them contain either a physical hazard, like an open hole in the ground, explosives or an environmental hazard, like mercury. By law, producing coal mines have to give money to the U.S. government for each ton of coal that is mined. The money is supposed to be used to clean up abandoned coal mines. However, there is no such law to require other mineral producers to give money to clean up abandoned non-coal mines. Therefore, States with predominantly non-coal abandoned mines, like Alaska, lobbied to tap into the abandoned mine fund in order to clean up the non-coal abandoned mines. Since most of the abandoned mines are on Federal land, it was only logical the Federal government should fund an effort to identify the abandoned mines and mitigate the hazards at those mines. The BOM was uniquely suited to conduct these inventories because it had a database with thousands of mine locations.

Because the BOM had located and mapped most of the abandoned mines for the Forest Service, we helped the Forest Service mitigate hazards on their land. In 1993 and 1994, I was involved with inventorying abandoned mines for the BLM in the Fortymile Wild and Scenic River corridor and Koyukuk Mining District. Unfortunately, I was not able to finish the investigations before leaving Alaska. The abandoned mine inventory work exemplified how far Alaska had changed. Alaska went from a land wide open to mineral development, where people did anything with little government oversight to a land where government entities were actively obliterating all signs of past mineral development.

Another responsibility I had as a supervisor was to give tours to people from the Department of Interior and BOM Washington Office. I always disliked interrupting the field routine by bringing people out for their annual junket to Alaska and also resented the fact these people might only come to Alaska because it was a free vacation. However, over time, my opinion of these trips mellowed and I understood how important it was to kowtow to the people who were in charge of the money. Probably the most memorable experience I had with the Washington Office bigwigs was when I took them on a tour of the Valdez Creek Mine, which is located on the Denali Highway. It was a deep open pit placer gold mine where the mine had to remove nearly 100 feet of gravel to get to bedrock. I brought the Washington Office guys to the mine when the mine had reached bedrock. While we were standing at the bottom of the pit, the mine manager allowed us to pick up the gold nuggets lying all around our feet. After a couple of minutes of picking, we gave the mine operator handfuls of gold nuggets. The operator admitted their mining method on bedrock was for a crew to walk across the bedrock picking up the gold by hand. We were all glad we were allowed to help the picking crew.

In the 1990's, things changed, not only for Alaska but for the BOM. Budgets got tighter and I was faced with trying to save jobs. This included closing the Fairbanks office, moving the permanent employees to Anchorage and not hiring any seasonal employees. Even though this led to more fieldwork for me because of the lack of personnel, it was a very depressing time. Times only got worse with the election of Bill Clinton as President in 1993. The Clinton administration seemed to favor environmental projects. It was hard to get any projects funded that featured mineral evaluations. The BOM tried to stay fully funded by pitching projects with environmental slants but the strategy didn't work. The BOM even tried saving itself by proposing to close all of the field offices and moving some select personnel to its Research Centers in the lower 48. They started the process by moving some of the people in the Alaska BOM offices to Research Centers in 1994. With the handwriting on the wall, in 1995, I transferred to the BLM National

Training Center in Arizona to teach people how to be field and government geologists. BOM discovered closing the field offices wasn't enough to save the agency. It was eventually defunded in 1996. Some of the people I worked with who had transferred to Research Centers were now out of jobs. Luckily the BOM personnel who remained in Anchorage were absorbed by BLM, thanks to Senator Ted Stevens. The former BOM people were able to continue doing some work for a number of years, but it was never at the same level as the good ol' BOM. Eventually, these people were absorbed into the BLM bureaucracy. A very important project completed by the BOM personnel for BLM was the digitization of all of the BOM reports on Alaska. These reports can be accessed on the ADGGS website. However, the bottom line was that because of the closure of the BOM, Alaska lost a lot of mineral-related expertise, mineral data and an advocate for the mineral industry.

Some final words about fieldwork in Alaska in the 1970s, 80s and 90s. Sometimes fieldwork was hard, lonely, dirty, sweaty, tiring, dangerous and for the most part fun. Where else would a person in their late 20's and early 30's get a budget of millions of dollars, a field crew, equipment, helicopters, boats, ATVs and told to go out and find something? I hope Alaska hasn't changed so much to prevent the exploitation of its mineral resources in an environmentally sound manner. Alaska needs mineral development for its economy and the U.S. needs Alaska's minerals for its security. Young geologists need somewhere to hunt for the treasures the earth contains and test their wilderness skills. For me, I'm now too old to live that life, but if I had the chance to do it all over again, I would say, "Sign me up, but first give me the body of a 20-year-old, with strong knees and a strong back."

If you like historical fiction, check out the *Life on the Alaska Frontier* series from Fleeting Edge Press:

Forging North

At the close of the 19th century, the gold rush is in full swing. A determined young man has left Seattle behind to head north. With the lure of Alaska gold burning bright in his eyes, Thomas Thornton set out on a voyage to find fame and fortune. He left behind all that he had, with a promise to his girl he would return. Thomas soon learned that Alaska had other ideas about his future. With the grit and determination demanded by those that seek to tame Alaska, Thomas vows to see his dreams come true.

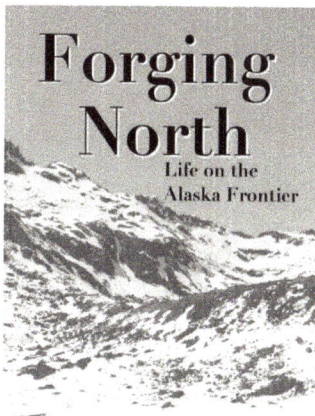

Fortymile

The promise of a rich claim in the Fortymile drove Thomas Thornton to risk it all on his dream of Alaska gold. Pushing north along the trail from Valdez to Chicken, the hardships are overshadowed by a mystery—one that may spell ruin and disaster. With the resolve of the early pioneers, he vows to overcome—to find the gold that will ensure his future. This sequel to "Forging North" follows Thomas as he seeks his fortune in the far north.

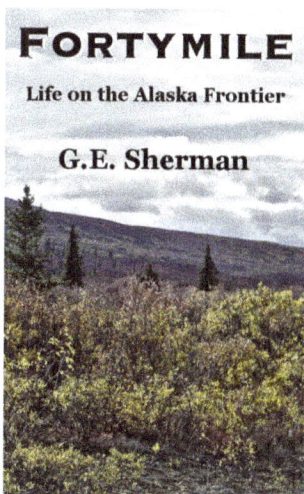

Available in Kindle and paperback on Amazon. See fleetingedgepress.com

www.ingramcontent.com/pod-product-compliance
Lightning Source LLC
Chambersburg PA
CBHW060542200326
41521CB00007B/455